Ernst Schering Research Foundation Workshop 31
Advances in Eicosanoid Research

Springer
*Berlin
Heidelberg
New York
Barcelona
Hong Kong
London
Milan
Paris
Singapore
Tokyo*

Ernst Schering Research Foundation
Workshop 31

# Advances in Eicosanoid Research

C. N. Serhan, H. D. Perez
Editors

With 51 Figures and 12 Tables

Springer

Series Editors: G. Stock and M. Lessl

ISSN 0947-6075
ISBN 3-540-66747-4   Springer-Verlag Berlin Heidelberg New York

CIP data applied for

Die Deutsche Bibliothek – CIP-Einheitsaufnahme
Schering-Forschungsgesellschaft <Berlin>: Ernst Schering Research Foundation Workshop. - Berlin; Heidelberg; New York; Barcelona; Budapest; Hong Kong; London; Milan; Paris; Santa Clara; Singapore; Tokyo: Springer.
ISSN 0947-6075
29. Advances in Eicosanoid Research. - 2000
  Advances in Eicosanoid Research ; with tables / C. N. Serhan and H. D. Perez, ed. - Berlin; Heidelberg; New York; Barcelona; Budapest; Hong Kong; London; Milan; Paris; Singapore; Tokyo: Springer, 2000
(Ernst Schering Research Foundation Workshop; 29)
ISBN 3-540- 66747-4

This work is subject to copyright. All rights are reserved, whether the whole or part of the material is concerned, specifically the rights of translation, reprinting, reuse of illustrations, recitation, broadcasting, reproduction on microfilms or in any other way, and storage in data banks. Duplication of this publication or parts thereof is permitted only under the provisions of the German Copyright Law of September 9, 1965, in its current version, and permission for use must always be obtained from Springer-Verlag. Violations are liable for prosecution under the German Copyright Law.

Springer-Verlag is a company in the BertelsmannSpringer publishing group.
© Springer-Verlag Berlin Heidelberg 2000
Printed in Germany

The use of general descriptive names, registered names, trademarks, etc. in this publication does not imply, even in the absence of a specific statement, that such names are exempt from the relevant protective laws and regulations and therefore free for general use. Product liability: The publishers cannot guarantee the accuracy of any information about dosage and application contained in this book. In every individual case the user must check such information by consulting the relevant literature.

Typesetting: Data conversion by Springer-Verlag
Printing: Druckhaus Beltz, Hemsbach. Binding: J. Schäffer GmbH & Co. KG, Grünstadt
SPIN:10751302     21/3134/AG–5 4 3 2 1 0 – Printed on acid-free paper

# *Preface*

Over the last few years, we have witnessed tremendous progress in the field of eicosanoids and their therapeutic applications. Receptor antagonists for leukotrienes have been tested as anti-inflammatories and are on the market as a treatment for asthma. Receptor agonists for prostacyclin are being tested for the treatment of peripheral vascular disease, and selective inhibitors of cyclooxygenase type II were just approved for the treatment of rheumatoid arthritis. All these developments are the culmination of many years and man-hours of careful research.

The field has now entered an upswing that will result in novel therapeutic applications within the next 10 years. New molecules and mediators have been identified, new enzymes and pathways elucidated and new therapeutic approaches have emerged. The concept of eicosanoids as "pro-inflammatory" molecules is being challenged, and their role as regulators is increasingly recognized. In fact, some of these molecules may be important endogenous anti-inflammatory agents.

This workshop was first conceived approximately 2 years ago, when we were reflecting on the history of the eicosanoid field, our experiences as graduate students and fellows and the tremendous change that this field has undergone since then. It was our intention to bring together an outstanding group of scientists spanning the broad scope of eicosanoid research and to have open discussions of future developments. The task was facilitated by the generous support provided by the Ernst Schering Research Foundation.

This volume contains the text of the main lectures presented at the workshop "Advances in Eicosanoid Research" held in San Francisco, California, from July 28 to July 30, 1999. Topics include mechanisms of action of anti-inflammatory drugs, enzymes involved in arachidonic-acid metabolism, their cloning and crystal structure, the effect of inhibitors and cell knockouts, novel receptors and their mechanism of signal transduction, anti-inflammatory eicosanoids and eicosanoid regulation of inflammation and metastasis.

Our understanding of the cellular and molecular basis of the inflammatory response continues to evolve with the discovery of new pathways and mediators. The quest for new anti-inflammatory agents depends on the elucidation of biochemical pathways involved in the initiation and termination of the inflammatory response. The topics presented here are state-of-the-art and provide insights into novel approaches. The editors trust that the reader will share our enthusiasm and continued excitement about progress in the field of eicosanoids.

*Charles N. Serhan, H. Daniel Perez*

# Contents

1  The Mechanism of Action of Anti-Inflammatory Drugs
   J.R. Vane . . . . . . . . . . . . . . . . . . . . . . . . . .  1

2  Human Group-V Phospholipase-A$_2$ Expression
   in *Pichia pastoris* and Its Role in Eicosanoid Generation
   L. J. Lefkowitz, H. Shinohara, E. A. Dennis . . . . . . . . .  25

3  Fatty-Acid Substrate Interactions with Cyclooxygenases
   W. L. Smith, C. J. Rieke, E. D. Thuresson, A. M. Mulichak,
   R. M. Garavito . . . . . . . . . . . . . . . . . . . . . . .  53

4  Structure, Function and Inhibition of Cyclooxygenases
   L. J. Marnett . . . . . . . . . . . . . . . . . . . . . . . .  65

5  Leukotriene-A$_4$ Hydrolase: Probing the Active Sites
   and Catalytic Mechanisms by Site-Directed Mutagenesis
   J. Z. Haeggström, A. Wetterholm . . . . . . . . . . . . . .  85

6  The Regulation of Cyclooxygenase-1 and -2
   in Knockout Cells and Cyclooxygenase and Fever
   in Knockout Mice
   L. R. Ballou . . . . . . . . . . . . . . . . . . . . . . . .  97

7  Leukotriene-B$_4$ Receptor and Signal Transduction
   T. Shimizu, T. Yokomizo, T. Izumi . . . . . . . . . . . . .  125

8   Lipoxins, Aspirin-Triggered 15-epi-Lipoxin Stable Analogs
    and Their Receptors in Anti-Inflammation:
    a Window for Therapeutic Opportunity
    *C. N. Serhan, B. D. Levy, C. Clish, K. Gronert, N. Chiang*  . 143

9   Lipoxin-Stable Analogs: Potential Therapeutic
    Downregulators of Intestinal Inflammation
    *A. T. Gewirtz, J. L. Madara* . . . . . . . . . . . . . . . . . . 187

10  The Role of Eicosanoids in Tumor Growth and Metastasis
    *D. Nie, K. Tang, K. Szekeres, M. Trikha, K. V. Honn* . . . . 201

Subject Index . . . . . . . . . . . . . . . . . . . . . . . . . . . . . . . . 219

Previous Volumes Published in this Series . . . . . . . . . . . . . . . . 223

# List of Editors and Contributors

## Editors

*C.N. Serhan*
Center for Experimental Therapeutics and Reperfusion Injury,
Department of Anesthesiology, Perioperative and Pain Medicine,
Brigham and Women's Hospital and Harvard Medical School,
75 Francis Street, Boston, MA 02115, USA

*H.D. Perez*
Dept. of Immunology, Berlex Biosciences, Richmond, CA 94804-0099, USA

## Contributors

*L.R. Ballou*
Departments of Medicine and Biochemistry, The University of Tennessee,
Memphis, TN 38163, USA

*N. Chiang*
Center For Experimental Therapeutics and Reperfusion Injury,
Department of Anesthesiology, Perioperative and Pain Medicine,
Brigham and Women's Hospital and Harvard Medical School,
75 Francis Street, Boston, MA 02115, USA

*C. Clish*
Center For Experimental Therapeutics and Reperfusion Injury,
Department of Anesthesiology, Perioperative and Pain Medicine,
Brigham and Women's Hospital and Harvard Medical School,
75 Francis Street, Boston, MA 02115, USA

*E.A. Dennis*
Department of Chemistry and Biochemistry, School of Medicine and Revelle College, University of California, San Diego, La Jolla, CA 92093-0601, USA

*R.M. Garavito*
513 Biochemistry Building, Department of Biochemistry, Michigan State University, East Lansing, MI 48824, USA

*A.T. Gewirtz*
Epithelial Pathobiology Unit, Department of Pathology and Laboratory Medicine, Emory University School of Medicine, Atlanta, GA 30322, USA

*K. Gronert*
Center For Experimental Therapeutics and Reperfusion Injury,
Department of Anesthesiology, Perioperative and Pain Medicine,
Brigham and Women's Hospital and Harvard Medical School,
75 Francis Street, Boston, MA 02115, USA

*J.Z. Haeggström*
Department of Medical Biochemistry and Biophysics,
Division of Chemistry II, Karolinska Institutet, S-171 77 Stockholm, Sweden

*K.V. Honn*
Department of Radiation Oncology, Wayne State University,
430 Chemistry Building, Detroit, MI 48202, USA

*T. Izumi*
Department of Biochemistry and Molecular Biology, Faculty of Medicine, University of Tokyo, Hongo 7-3-1, Bunkyo-ku, Tokyo 113-0033, Japan

*L.J. Lefkowitz*
Department of Chemistry and Biochemistry, School of Medicine and Revelle College, University of California San Diego, La Jolla, CA 92093-0601, USA

*B.D. Levy*
Center For Experimental Therapeutics and Reperfusion Injury,
Department of Anesthesiology, Perioperative and Pain Medicine,
Brigham and Women's Hospital and Harvard Medical School,
75 Francis Street, Boston, MA 02115, USA

# List of Editors and Contributors

*J.L. Madara*
Epithelial Pathobiology Unit, Department of Pathology and Laboratory
Medicine, Emory University School of Medicine, Atlanta, GA 30322, USA

*L.J. Marnett*
Department of Biochemistry, Vanderbilt University School of Medicine,
Nashville TN 37232, USA

*A.M. Mulichak*
513 Biochemistry Building, Department of Biochemistry,
Michigan State University, East Lansing, MI 48824, USA

*D. Nie*
Department of Radiation Oncology, Wayne State University,
430 Chemistry Building, Detroit, MI 48202, USA

*C.J. Rieke*
513 Biochemistry Building, Department of Biochemistry,
Michigan State University, East Lansing, MI 48824, USA

*C.N. Serhan*
Center For Experimental Therapeutics and Reperfusion Injury,
Department of Anesthesiology, Perioperative and Pain Medicine,
Brigham and Women's Hospital and Harvard Medical School,
75 Francis Street, Boston, MA 02115, USA

*T. Shimizu*
Department of Biochemistry and Molecular Biology, Faculty of Medicine,
University of Tokyo, Hongo 7-3-1, Bunkyo-ku, Tokyo 113-0033, Japan

*H. Shinohara*
Department of Chemistry and Biochemistry, School of Medicine and Revelle
College, University of California San Diego, La Jolla, CA 92093-0601, USA

*W.L. Smith*
513 Biochemistry Building, Department of Biochemistry,
Michigan State University, East Lansing, MI 48824, USA

*K. Szekeres*
Department of Radiation Oncology, Wayne State University,
430 Chemistry Building, Detroit, MI 48202, USA

*K. Tang*
Department of Radiation Oncology, Wayne State University,
430 Chemistry Building, Detroit, MI 48202, USA

*E.D. Thuresson*
513 Biochemistry Building, Department of Biochemistry,
Michigan State University, East Lansing, MI 48824, USA

*M. Trikha*
Department of Radiation Oncology, Wayne State University,
430 Chemistry Building, Detroit, MI 48202, USA

*J.R. Vane*
The William Harvey Research Institute, St. Bartholomew's
and Royal London School of Medicine and Dentistry, Charterhouse Square,
London, EC1M 6BQ, UK

*A. Wetterholm*
Department of Medical Biochemistry and Biophysics,
Division of Chemistry II, Karolinska Institutet, S-171 77 Stockholm, Sweden

*T. Yokomizo*
Department of Biochemistry and Molecular Biology, Faculty of Medicine,
University of Tokyo, Hongo 7-3-1, Bunkyo-ku, Tokyo 113-0033, Japan

# 1 The Mechanism of Action of Anti-Inflammatory Drugs

J. R. Vane

| | | |
|---|---|---:|
| 1.1 | Historical Introduction | 1 |
| 1.2 | Properties of COX-1 and COX-2 | 4 |
| 1.3 | Functions of COX-1 and COX-2 | 6 |
| 1.4 | Selective Inhibition of COX-2 | 8 |
| 1.5 | Conclusions | 15 |
| References | | 16 |

## 1.1 Historical Introduction

The forefather of aspirin, salicylic acid, is a constituent of several plants long used as medicaments. About 3500 years ago, the Ebers papyrus recommended the application of a decoction of dried leaves of myrtle to the abdomen and back to expel rheumatic pains from the womb. A thousand years later, Hippocrates championed the juices of the poplar tree for treating eye diseases and those of willow bark for pain in childbirth and for fever. All contain salicylate.

Celsus (in AD 30) described the four famous signs of inflammation (rubor, calor, dolor and tumor, or redness, heat, pain and swelling) and used extracts of willow leaves to relieve them. Through Roman times, the use of salicylate-containing plants was further developed, and willow bark was recommended for mild to moderate pain. Salicylate-containing plants were also being applied in Asia and China. The North American Indians and the Hottentots of Southern Africa knew the curative effects of *Salix* and *Spirea* species. In the Middle Ages, further uses were found: plasters used to treat wounds and various other external and internal applications, including the treatment of menstruation and dysentery.

A country parson, the Reverend Edward Stone of Chipping Norton in Oxfordshire, made the first "clinical trial" of Willow bark. He wrote a letter on the use of willow bark in fever dated June 2, 1763 to the President of the Royal Society of London (Stone 1763). He took a pound of willow bark, dried it on a baker's oven for 3 months and pulverized it. He described his success with doses of 1 dram in about 50 febrile patients.

He concluded his paper by saying, "I have no other motives for publishing this valuable specific, than that it may have a fair and full trial in all its variety of circumstances and situations, and that the world may reap the benefits occurring from it". His wishes have certainly been realized; world consumption of aspirin is estimated at 45 000 tons per year, with an average consumption in a developed country of about 100 tablets per person per year. Without the discovery of many replacements for aspirin (the non-steroid anti-inflammatory drugs or NSAIDs) in recent years, consumption would surely have been much higher.

Salicylic acid was synthesized in 1874 by Kolbe in Germany. MacLagan (1876) and Stricker (1876) showed that salicylic acid was effective in rheumatic fever and, a few years later, in the 1880s, sodium salicylate was also in use as a treatment for chronic rheumatoid arthritis and gout and as an antiseptic compound. Felix Hoffman was a young chemist working at Bayer. His father, who had been taking salicylic acid for many years to treat his arthritis, complained strenuously to his son about its bitter taste. Felix responded by adding an acetyl group to salicylic acid to make acetylsalicylic acid.

Heinrich Dreser, as the head of the first industrial pharmacology laboratory set up by Bayer in 1890, showed acetylsalicylic acid to be analgesic, antipyretic and anti-inflammatory. Bayer introduced the new drug as "aspirin" in 1899, and sales have increased ever since (Dreser 1899).

Several other NSAIDs that shared some or all of these effects were discovered. Among these were antipyrine, phenacetin, phenylbutazone and, more recently, the fenamates, indomethacin and naproxen.

Despite the diversity of their chemical structures, these drugs all share the same therapeutic properties. In varying doses, they alleviate the swelling, redness and pain of inflammation, reduce general fever and cure headaches. They also share (to a greater or lesser extent) a number of similar side effects. They can cause gastric ulcer, delay the birth process and damage the kidney. A particularly interesting "side effect", now known as a therapeutic action, is the anti-thrombotic effect. Many clinical trials have shown that aspirin once daily in doses as low as 75 mg helps to prevent heart attacks and strokes.

When a chemically diverse group of drugs all share the same therapeutic qualities (which in themselves have not much connection with each other) and the same side effects, it is likely that the actions of those drugs are based on a single biochemical intervention. For many years, pharmacologists and biochemists searched for such a common mode of action without finding a generally acceptable scientific explanation.

Before 1971, many biochemical effects of the NSAIDs had been documented, and there were many hypotheses about their mode of action. They included uncoupling of oxidative phosphorylation and inhibition of dehydrogenase enzymes and key enzymes involved in protein and RNA biosynthesis. However, the concentration of the drugs required for enzyme inhibition was in excess of the concentrations typically found in plasma after therapy, and there was no convincing reason why inhibition of any of these enzymes should produce the triple anti-inflammatory, analgesic and antipyretic effects of aspirin.

It was against this background of knowledge that the investigation of the mode of action of NSAIDs was begun by prostaglandin researchers. Piper and Vane (1969) described the anaphylactic release of prostaglandins and of another, very ephemeral, substance that, was named "rabbit aorta-contracting substance" (RCS) from isolated, perfused lungs of the guinea pig. In the lung perfusate, RCS had a half-life of about 2 min; Samuelsson's group (Hamberg et al. 1975) identified it as thromboxane A2 in 1975. Palmer et al. (1970) found that aspirin and similar drugs blocked the release of RCS from guinea pig isolated lungs during anaphylaxis. Vane postulated that the various stimuli, which released prostaglandins, were "turning on" the production of these compounds and that aspirin might be blocking their synthesis. Using the supernatant of a broken cell homogenate from guinea pig lung as a source of prostaglandin synthase, Vane (1971) found a dose-dependent inhibition of prostaglandin formation by aspirin, salicylate and indomethacin but not by morphine. Two other reports from the same laboratory lent support to and extended this finding. Smith and Willis (1971) found that aspirin prevented the release of prostaglandins from aggregating human platelets, and Ferreira et al. (1971) demonstrated that aspirin-like drugs blocked prostaglandin release from the perfused, isolated spleen of the dog.

The discovery that all NSAIDs act by inhibiting the enzyme we now call cyclo-oxygenase (COX), which leads to the generation of prostaglandins, provided a unifying explanation for their therapeutic actions and firmly established certain prostaglandins as important mediators of inflammatory disease (Vane and Botting 1997; Vane et al. 1998). Hemler et al. (1976) isolated COX (or prostaglandin endoperoxide synthase, PGHS) with a molecular weight of 71 kDa, and DeWitt and Smith, Yokoyama et al. and Merlie et al. cloned it in 1988. The enzyme exhibits both COX and hydroperoxidase activities. COX first cyclizes arachidonic acid to form prostaglandin $G_2$, and the peroxidase then reduces prostaglandin $G_2$ to prostaglandin $H_2$.

## 1.2 Properties of COX-1 and COX-2

For the last 10 years, we have known that COX exists as two isoforms, COX-1 and COX-2. Garavito and his colleagues (Picot et al. 1994) have determined the three-dimensional structure of COX-1, which consists of three independent folding units: an epidermal growth factor-like domain, a membrane-binding section and an enzymatic domain. The sites for peroxidase and COX activity are adjacent but spatially distinct. The enzyme integrates into only a single leaflet of the membrane lipid bilayer and, thus, the position of the COX channel allows arachidonic acid to gain access to the active site from the interior of the bilayer. Most NSAIDs compete with arachidonic acid for binding to the active site. Uniquely, aspirin irreversibly inhibits COX-1 by acetylation of serine 530, thereby excluding access for the substrate (Roth et al. 1975). The three-dimensional structure of COX-2 (Luong et al. 1996) closely resembles that of COX-1 but, fortunately for the medicinal chemist, the binding sites for arachidonic acid on these enzymes are slightly different. The active site of COX-2 is slightly larger and can accommodate bigger structures than those that are able to reach the active site of COX-1. A secondary internal pocket contributes significantly to the larger volume of the active site of COX-2, although the central channel is also bigger by 17%.

The constitutive isoform, COX-1, has clear physiological functions. Its activation leads, for instance, to the production of prostacyclin which, when released by the endothelium, is anti-thrombogenic (Moncada et al. 1976) and, when released by the gastric mucosa, is cytoprotective (Whittle et al. 1980). The inducible isoform, COX-2, is induced in a number of cells by pro-inflammatory stimuli (Xie et al. 1992). Its existence was first suspected when Needleman and his group (Fu et al. 1990) reported that bacterial lipopolysaccharide (LPS) increased the synthesis of prostaglandins in human monocytes in vitro and in mouse peritoneal macrophages in vivo (Masferrer et al. 1990). This increase was inhibited by dexamethasone and associated with de novo synthesis of new COX protein. A year or so later, an inducible COX was identified as a distinct isoform of COX (COX-2) encoded by a different gene from that encoding COX-1 (Kujubu et al. 1991; O'Banion et al. 1991; Xie et al. 1991; Sirois and Richards 1992). Both enzymes have a molecular weight of 71 kDa, and the amino acid sequence of the complementary DNA for COX-2 shows a 60% homology with the sequence of the non-inducible enzyme. The messenger RNA (mRNA) for the inducible enzyme is approximately 4.5 kb, and that of the constitutive enzyme is 2.8 kb. The inhibition of the expression of COX-2 by glucocorticoids is an additional aspect of the anti-inflammatory action of corticosteroids. The levels of COX-2, normally very low in cells, are tightly controlled by a number of factors, including cytokines, intracellular messengers and the availability of substrate.

Since COX-2 is induced by inflammatory stimuli and by cytokines in migratory and other cells, it is attractive to suggest that the anti-inflammatory actions of NSAIDs are due to the inhibition of COX-2. Likewise, the unwanted side effects, such as damage to the stomach lining, are due to inhibition of the constitutive enzyme, COX-1. This unifying concept is now becoming generally accepted.

## 1.3 Functions of COX-1 and COX-2

### 1.3.1 Gastrointestinal Tract

The so-called "cytoprotective" action of prostaglandins in preventing gastric erosions and ulceration is mainly brought about by endogenously produced prostacyclin and prostaglandin $E_2$ ($PGE_2$), which reduce gastric-acid secretion and exert a direct vasodilator action on the vessels of the gastric mucosa. In addition to these major actions, Whittle and Vane (1987) showed that prostanoids stimulate the secretion of viscous mucus and duodenal bicarbonate but inhibit gastric fluid. In most species, including humans, the protective prostaglandins are synthesized by COX-1, although Kargman et al. (1996) found small quantities of COX-2 in the normal rat stomach. COX-2 is expressed in human gastric mucosa infected with *Helicobacter pylori* (Jackson et al. 1998) and in ulcerative colitis (McLaughan et al. 1996). COX-2 is also expressed around the periphery of gastric ulcers in mice and rats. It may be involved in wound healing, because COX-2 inhibitors delay ulcer healing (Mizuno et al. 1997; Schassmann et al. 1998). Large quantities of COX-2 are expressed in experimentally induced and human colon cancers (Gustafson-Svärd et al. 1996; Kutchera et al. 1996)

It is at first puzzling why knockout mice in which the COX-1 gene has been deleted (Langenbach et al. 1995) do not develop gastric ulcers spontaneously but do show a decreased sensitivity to the damaging effects of indomethacin. However, the normality of the mucosa in these mice could well be brought about by the continued release of nitric oxide and calcitonin gene-related protein, both also known to contribute to the maintenance of healthy mucosa (Whittle 1993).

### 1.3.2 Kidney

Maintenance of kidney function both in animal models of disease states and in patients with congestive heart failure, liver cirrhosis or renal insufficiency, is dependent on vasodilator prostaglandins. These patients are, therefore, at risk of renal ischemia when prostaglandin synthesis is reduced by NSAIDs. Synthesis of $PGE_2$ and prostacyclin takes place mainly via COX-1, though Harris et al. (1994) have reported low levels of mRNA for COX-2. There is also up-regulation of COX-2 expression in the macula densa following salt deprivation. COX-2 expression may drive the renin–angiotensin system (Harris 1996). Schneider and Stahl (1998) have reviewed this rapidly evolving field.

Mice that lack the gene for production of COX-1 appear to be healthy and do not show significant signs of kidney pathology. This is in accord with the finding that inhibition of COX-1 by NSAIDs does not alter renal function under normal physiological conditions. However, in COX-2 (–/–) null mice, the kidneys failed to develop fully after birth, with the result that the animals died before they were 8 weeks old (Morham et al. 1995).

### 1.3.3 Central Nervous System

COX-1 is found in neurons throughout the brain, but it is most abundant in the forebrain, where prostaglandins may be involved in complex integrative functions, such as control of the autonomic nervous system, and in sensory processing. COX-2 mRNA is induced in brain tissue and in cultured glial cells by pyrogenic substances, such as LPS, interleukin 1 (IL-1) or tumor necrosis factor (Cao et al. 1998). However, low levels of COX-2 protein and COX-2 mRNA have been detected in neurons of the forebrain without previous stimulation by pro-inflammatory stimuli. These "basal" levels of COX-2 are particularly high in neonates and are probably induced by physiological nervous activity. Intense nerve stimulation, leading to seizures, induces COX-2 mRNA in discrete neurons of the hippocampus (Marcheselli and Bazan 1996), whereas acute stress raises levels in the cerebral cortex. COX-2 mRNA is also constitutively expressed in the spinal cord of normal rats and may be involved with processing of nociceptive stimuli (Beiche et al. 1996). Endogenous, fever-producing $PGE_2$ is thought to originate from COX-2 induced by LPS or IL-1 in endothelial cells lining the blood vessels of the hypothalamus (Cao et al. 1998). The specific COX-2 inhibitor rofecoxib is a potent antipyretic agent in man (Schwartz et al. 1998).

### 1.3.4 Reproductive System

Expression of COX-1 is much greater than that of COX-2 in fetal hearts, kidneys, lungs and brains and in the decidual lining of the uterus (Bennett and Slater 1996; Gibb and Sun 1996). Prostaglandins synthesized by COX-1 are apparently essential for the survival of fetuses during parturition, since the majority of offspring born to homozygous COX-1 knockout mice do not survive (Langenbach et al. 1995). The high mortality of the pups may be due to premature closure of the ductus arteriosus. Female COX-2 knockout mice are mostly infertile, producing very few offspring, due to a reduction in ovulation (Dinchuck et al. 1995). COX-2 induction is involved in ovulation and is clearly the trigger for parturition (Brown et al. 1998; Slater et al. 1998), leading to $PGF_{2\alpha}$ release, which causes contractions of the uterine smooth muscle.

### 1.3.5 Endothelium

Endothelial cells generate prostacyclin, which is anti-aggregatory and is a vasodilator. They clearly contain COX-1, but McAdam et al. (1999), using the specific COX-2 inhibitor celecoxib, found that urinary excretion of the prostacyclin metabolite was substantially suppressed by celecoxib at doses up to 800 mg in volunteers. They concluded that COX-2 is induced in endothelial cells in the circulation, probably by laminar shear stress (Topper et al. 1996).

## 1.4 Selective Inhibition of COX-2

Sales of anti-inflammatory drugs are now estimated at 5.8 billion dollars per year, with the USA accounting for 1.8 billion dollars. However, they all have toxic effects on the stomach. Indeed, several epidemiological studies have characterized the degree of gastric damage caused by different compounds (Henry et al. 1996). An estimated 34–46% of patients receiving NSAID therapy will have some form of adverse gastrointestinal event. In the USA alone, some 100,000 patients taking NSAIDs are hospitalized each year because of perforations, ulcers or bleeding in the stomach (PUBs; Fries 1996). Some 15% of these patients die in intensive care. Of course, these hospitalizations only represent extreme gastric irritation, which ranges from mild dyspepsia to PUBS. Even ibuprofen, recognized as one of the mildest gastric irritants, causes problems in a significant proportion of patients. Clearly, there is dramatic need for anti-inflammatory drugs that do not affect the stomach.

### 1.4.1 COX-2/COX-1 Ratios

The importance of the discovery of the inducible COX-2 is highlighted by the differences in pharmacology of the two enzymes (Mitchell et al. 1993). Aspirin, indomethacin and ibuprofen are much less active against COX-2 than against COX-1 (Meade at al. 1993). Indeed, the strongest inhibitors of COX-1, such as aspirin, indomethacin and piroxicam, are the NSAIDs that cause the most damage to the stomach (Lanza 1989). The spectrum of activities of some ten standard NSAIDs against the two enzymes ranges from a high selectivity toward COX-1 (166-fold for aspirin) to equiactivity for both enzymes (Akarasereenont et al. 1994).

There are now many methods for measuring COX-2/COX-1 ratios, varying from human enzyme assays in vitro (Churchill 1996) to assays in human blood samples (Patrignani et al. 1994; Warner et al. 1999; Giuliano and Warner 1999). In general, the whole-blood assays are reproducible from laboratory to laboratory and provide conditions that approach physiological conditions. For instance, any plasma protein binding will automatically be taken into account. Table 1 compares the COX-2/COX-1 ratios found by different authors.

NSAIDs are used in inflammation because they inhibit COX-2. The range of activities of NSAIDs against COX-1 compared with COX-2 explains the variations in the side effects of NSAIDs at their anti-inflammatory doses. Drugs that have a high potency against COX-2 and a low COX-2/COX-1 activity ratio have potent anti-inflammatory activity with few side effects on the stomach and kidney. Garcia-Rodriguez and Jick (1994) have published a comparison of epidemiological data on the side effects of NSAIDs. Piroxicam and indomethacin in anti-inflammatory doses showed high gastrointestinal toxicity. These drugs have a much higher potency against COX-1 than against COX-2 (Vane and Botting 1997). Thus, when epidemiological results are compared with

**Table 1.** The Cox-2/Cox-1 ratios of the $ED_{50}$ values for some non-steroidal anti-inflammatory drugs and selective/specific cyclooxygenase-2 inhibitors in whole-blood assays in different laboratories

| Drug | Warner et al. 1999a | Warner et al. 1999[b] | Patrignani et al. 1996 | Brideau et al. 1996 | Pairet and Van Ryn 1998 | Glaser 1995 |
|---|---|---|---|---|---|---|
| Ketoprofen | 5.1 | 61 | 1.7 | 5.4 | | |
| Flurbiprofen | 10.0 | 73 | 1.0 | 14.6 | | |
| Indomethacin | 10.0 | 80 | 0.53 | 2.88 | 0.82 | 5.7 |
| Piroxicam | 0.1 | 3.3 | 0.32 | 11.8 | 1.1 | |
| Naproxen | 3.8 | 3.0 | 1.67 | 9.5 | | 13.1 |
| Ibuprofen | 2.6 | 0.9 | 2.0 | 6.3 | | |
| 6-MNA | 2.6 | > 5 | 0.67 | | | |
| Diclofenac | 0.3 | 0.5 | | 0.36 | 0.39 | 1.5 |
| Etodolac | 0.1 | 0.2 | | | | 0.09 |
| Nimesulide | 0.038 | 0.19 | 0.006 | | | |
| Meloxicam | 0.04 | 0.37 | 0.009 | | 0.08 | |
| Celecoxib | 0.3 | 0.7 | | | 0.029 | |
| NS-398 | 0.0061 | 0.051 | 0.006 | 0.09 | | 0.00003 |
| SC-58125 | | < 0.01 | 0.007 | < 0.033 | 0.027 | < 0.001 |
| L-745,337 | .01 | < 0.01 | 0.007 | < 0.3 | | |
| Rofecoxib | 0.0049 | 0.013 | | | | |

*MNA*, methoxy-2-naphthylacetic acid.
[a]Modified whole-blood assay.
[b]Unmodified whole-blood assay.

COX-2/COX-1 ratios, there is a parallel relationship between gastrointestinal side effects and COX-2/COX-1 ratios.

### 1.4.2 Selective COX-2 Inhibitors in Current Therapeutic Use

Meloxicam, nimesulide and etodolac were identified in the 1980s as potent anti-inflammatory drugs with low ulcerogenic activities in rat stomach. In some instances, this was also shown to parallel low activity against prostaglandin synthesis in rat stomach. After the characterization of the COX-2 gene, these three drugs were each found to preferentially inhibit COX-2 rather than COX-1, with a variation in their COX-2/COX-1 ratios of between 0.1 and 0.01, depending on the test system used. Ratios obtained by the human whole-blood assay, which measures inhibition of COX-1 in platelets and COX-2 in mononuclear cells stimulated with LPS, are now generally accepted as the best reflection of the inhibitory activity of the drugs in humans (Table 1).

Meloxicam, which has a selectivity toward COX-2 of between three- and tenfold in human whole-blood assays and of 100-fold in human recombinant enzymes (Churchill et al. 1996), is marketed around the world for use in rheumatoid arthritis and osteoarthritis. In double-blind trials (Dequeker et al. 1998; Hawkey et al. 1998) in many thousands of patients with osteoarthritis, meloxicam (in doses of 7.5 mg or 15 mg once daily) was comparable in efficacy with standard NSAIDs, such as naproxen (750–1000 mg), piroxicam (20 mg) or diclofenac (100 mg). Both doses of meloxicam produced significantly fewer adverse gastrointestinal effects than the standard NSAIDs ($P<0.05$). Discontinuation of treatment due to gastrointestinal side effects was also significantly less frequent with meloxicam. Perforations, ulcerations and bleeding occurred in fewer meloxicam-treated patients than in patients treated with piroxicam, diclofenac or naproxen. The frequency of adverse events with meloxicam was significantly less (at $P<0.05$) when compared with piroxicam or naproxen (Barner 1996; Distel et al. 1996).

Etodolac is marketed in Europe and North America for the treatment of osteoarthritis and rheumatoid arthritis. It has about fivefold selectivity for COX-2 in human whole blood (Glaser 1995). In healthy human volunteers, etodolac twice daily did not suppress gastric-mucosal prostaglandin production and caused less gastric damage than naproxen (Laine et al. 1995). Patients with osteoarthritis or rheumatoid arthritis obtained relief from symptoms equal to that experienced when taking other commonly used NSAIDs with etodolac, but with a lower incidence of serious gastrointestinal toxicity (Cummings and Amadio 1994).

Nimesulide is currently sold in several European countries and in South America for the relief of pain associated with inflammatory conditions. It is a preferential inhibitor of COX-2, with about fivefold greater potency against this enzyme than against COX-1 in human whole-blood assays (Table 1). In limited clinical trials for its use in acute and chronic inflammation, it was more effective than placebo and had anti-inflammatory activity comparable to those of established NSAIDs. Interestingly, nimesulide seems safe to use in aspirin-sensitive asthmatics. Several recent studies in NSAID-intolerant asthmatic patients demonstrated that therapeutic doses of nimesulide did not induce asthmatic attacks, while high doses of 400 mg only precipitated mild asthma in 10% of patients (Senna et al. 1996). Perhaps aspirin-induced asthma is associated with COX-1 inhibition?

A recent epidemiological study (Garcia-Rodriguez et al. 1998) identified 1505 patients with upper gastrointestinal-tract bleeding. It showed nimesulide to cause a relative risk similar to that for naproxen (4.4 times control) and more than that for diclofenac (2.7 times control). Clearly, other factors are also involved, such as frequency of dosage, etc. As with other NSAIDs, nimesulide is used in different dosages and, when these are separated, the higher doses cause a much higher relative risk.

### 1.4.3 "Specific" COX-2 Inhibitors

Both Merck and Searle refer to their new COX-2 inhibitors as "specific", arguing that, at therapeutic doses, there is only inhibition of COX-2 and not of COX-1. A pharmacologist uses the word "specific" far more rigorously but, until proven otherwise in clinics, we shall accept this usage (Ford-Hutchinson 1998).

Needleman and his group at Monsanto/Searle have made inhibitors that are some 1000-fold more potent against COX-2 than against COX-1 in enzyme assays (Isakson et al. 1998). One of these, SC-58635 (celecoxib), is an effective analgesic for moderate to severe pain following tooth extraction (Hubbard et al. 1996). Celecoxib given to human volunteers for 7 days provided no evidence of gastric damage (Lanza et al. 1997a). Interestingly, in our whole-blood assay (Warner et al. 1999), celecoxib is only tenfold more active against COX-2 than against COX-1. Celecoxib was successfully launched in the USA in December 1998, but only with indications for osteoarthritis and rheumatoid arthritis. Extensive clinical trial results have not been published. Nevertheless, it is already selling at the rate of one billion dollars per year. Interestingly, although doses of up to 800 mg of celecoxib do not affect platelet aggregation in volunteers, there was an inhibition of serum thromboxane-B2 production of up to 70%, albeit with a shallow dose/response curve (McAdam et al. 1999)

Another specific COX-2 inhibitor from Merck, MK966 (rofecoxib or Vioxx) is currently approaching the market. In phase-I studies, a single dose of 250 mg daily for 7 days (which is ten times the anti-inflammatory dose) produced no adverse effects on the stomach mucosa, as evidenced by gastroscopy (Lanza et al. 1997b). After a single dose of 1 g, there was no evidence of COX-1 inhibition in platelets, but the activity of COX-2 in LPS-stimulated monocytes ex vivo was reduced. For postoperative dental pain, rofecoxib at 25–500 mg demonstrated analgesic activity equal to that of ibuprofen (Ehrich et al. 1999) and provided relief from symptoms in a 6-week study of osteoarthritis (Ehrich et al. 1997). Rofecoxib is also effective at 50 mg once daily in treating the pain caused by dysmenorrhea.

### 1.4.4 Future Therapeutic Uses for Selective COX-2 Inhibitors

#### *1.4.4.1 Premature Labor*
Prostaglandins induce uterine contractions during labor. NSAIDs, such as indomethacin, delay premature labor by inhibiting the production of prostaglandins but will also cause early closure of the ductus arteriosus and reduce urine production by the fetal kidneys (Sawdy et al. 1997). The delay in the birth process is most likely due to inhibition of COX-2, because mRNA for COX-2 increases substantially in the amnion and placenta immediately before and after the start of labor (Gibb and Sun 1996), whereas the side effects on the fetus are due to inhibition of COX-1. One cause of pre-term labor could be an intra-uterine infection resulting in release of endogenous factors that increase prostaglandin production by up-regulating COX-2 (Spaziani et al. 1996). Nimesulide reduces prostaglandin synthesis in isolated fetal membranes and has been used successfully to delay premature labor for a prolonged period without manifesting the side effects of indomethacin on the fetus (Sawdy et al. 1997).

#### *1.4.4.2 Colon Cancer*
Epidemiological studies have established a strong link between ingestion of aspirin and a reduced risk of developing colon cancer (Thun et al. 1991; Luk 1996). Sulindac also caused reduction of prostaglandin synthesis and regression of adenomatous polyps in 11 out of 15 patients with familial adenomatous polyposis (FAP), a condition in which many colorectal polyps develop spontaneously, with eventual progression to tumors. This indication that COX activity is involved in the process leading to colon cancer is supported by the demonstration that COX-2

(but not COX-1) is highly expressed in human and animal colon cancer cells and in human colorectal adenocarcinomas (Gustafson-Svard et al. 1996; Kutchera et al. 1996). Further support for the close connection between COX-2 and colon cancer has come from studies in the mutant *Apc* mouse, which is a model of FAP in humans. The spontaneous development of intestinal polyposis in these mice was strongly reduced either by deletion of the COX-2 gene or by treatment with a highly selective COX-2 inhibitor (Eberhart et al. 1994; Oshima et al. 1996; Sheng et al. 1997). Nimesulide also reduced the number and size of intestinal polyps in *Min* mice (Nakatsugi et al. 1997). The development of azoxymethane-induced colon tumors over a year was inhibited in celecoxib-fed rats (Kawamori et al. 1998). Thus, it is highly likely that COX-2 inhibitors could be used prophylactically to prevent colon cancer in genetically susceptible individuals without causing gastrointestinal damage.

### *1.4.4.3 Alzheimer's Disease*
The connection between COX and Alzheimer's disease has been based mostly on epidemiology because of the lack of an animal model of the disease. A number of studies have shown a significantly reduced odds ratio for Alzheimer's disease in those taking NSAIDs as anti-inflammatory therapy (McGeer and McGeer 1995; Breitner 1996; Cochran and Vitek 1996). The Baltimore Longitudinal Study of Aging (Stewart et al. 1997), with 1686 participants, showed that the risk of developing Alzheimer's disease is reduced among users of NSAIDs, especially those who have taken the medication for 2 years or more. No decreased risk was evident with acetaminophen or aspirin use. However, aspirin was probably taken in a dose too low to have an anti-inflammatory effect. The protective effect of NSAIDs is consistent with evidence of inflammatory activity in the pathophysiology of Alzheimer's disease. There is a strong interest in COX-2 in Alzheimer's disease, and Pasinetti and Aisen (1998) have shown expression of COX-2 in the frontal cortex of the brains of Alzheimer's patients.

## 1.5 Conclusions

The identification of selective inhibitors of COX-2 will clearly provide important advances in the therapy of inflammation. Conventional NSAIDs lead to gastrointestinal side effects (which include ulceration of the stomach, sometimes with subsequent perforation) and an estimated several thousand deaths per year in the USA alone. The evidence is strong, both from animal tests and from clinics, that selective COX-2 inhibitors will have greatly reduced side effects.

All the results so far published (and many reported at meetings, but yet to be fully published in the literature) support the hypothesis that the unwanted side effects of NSAIDs are due to their ability to inhibit COX-1, while their anti-inflammatory (therapeutic) effects are due to inhibition of COX-2. This concept is now certain. Thus, selective COX-2 inhibitors will provide an important advance in anti-inflammatory therapy. They are unlikely to be more potent anti-inflammatory agents than the conventional NSAIDs, but they will have the tremendous advantage of being safer and better tolerated. The clinical results with meloxicam already show this improved safety and tolerability, even though it retains some activity against COX-1. In the future, in addition to their beneficial actions in inflammatory diseases, these drugs may be useful for the prevention of colon cancer, Alzheimer's disease or premature labor.

Finally, the suppression of prostacyclin release from endothelial cells by specific COX-2 inhibitors suggests the possibility of interference with the cardiovascular system. However, we have been using COX-2 inhibitors for many years, because this is how NSAIDs produce their therapeutic effects. Thus, the specific inhibitors affect prostacyclin production just as conventional NSAIDs do. New side effects of the specific COX-2 inhibitors may arise from the fact that they cross the blood–brain barrier, whereas NSAIDs generally do not.

## References

Akarasereenont P, Mitchell JA, Thiemermann C, Vane JR (1994) Relative potency of nonsteroid anti-inflammatory drugs as inhibitors of cyclooxygenase-1 or cyclooxygenase-2. Br J Pharmacol 112[suppl]:183P

Barner A (1996) Review of clinical trials and benefit/risk ratio of meloxicam. Scand J Rheumatol Suppl 102:29–37

Beiche F, Scheuerer S, Brune K, Geisslinger G, Goppelt-Struebe M (1996) Up-regulation of cyclooxygenase-2 mRNA in the rat spinal cord following peripheral inflammation. FEBS Lett 390:165–169

Bennett P, Slater D (1996) COX-2 expression in labor. In: Vane J, Botting J, Botting R (eds) Improved non-steroid anti-inflammatory drugs. COX-2 enzyme inhibitors. Kluwer Academic, Lancaster, pp 167–188

Breitner JCS (1996) The role of anti-inflammatory drugs in the prevention and treatment of Alzheimer's disease. Annu Rev Med 47:401–411

Brideau C, Kargman S, Liu S, Dallob AL, Ehrich EW, Rodger IW, et al. (1996) A human whole blood assay for clinical evaluation of biochemical efficacy of cyclooxygenase inhibitors. Inflamm Res 45:68–74

Brown NL, Alvi SA, Elder MG, Bennett PR, Sullivan MHF (1998) A spontaneous induction of fetal membrane prostaglandin production precedes clinical labor. J Endocrinol 157:R1–R6

Cao C, Matsumura K, Yamagata K, Watanabe Y (1998) Cyclooxygenase-2 is induced in brain blood vessels during fever evoked by peripheral or central administration of tumor necrosis factor. Mol Brain Res 56:45–56

Churchill L, Graham AG, Shih C-K, Pauletti D, Farina PR, Grob PM (1996) Selective inhibition of human cyclo-oxygenase-2 by meloxicam. Inflammopharmacology 4:125–135

Cochran FR, Vitek MP (1996) Neuroinflammatory mechanisms in Alzheimer's disease: new opportunities for drug discovery. Expert Opin Invest Drugs 5:449–455

Cummings DM, Amadio P Jr (1994) A review of selected newer nonsteroidal anti-inflammatory drugs. Am Fam Physician 49:1197–1202

Dequeker J, Hawkey C, Kahan A, et al. (1998) Improvement in gastrointestinal tolerability of the selective cyclooxygenase (COX)-2 inhibitor, meloxicam, compared with piroxicam: results of the safety and efficacy large-scale evaluation of COX-inhibiting therapies (SELECT) trial in osteoarthritis. Br J Pharmacol 37:946–951

DeWitt DL, Smith WL (1988) Primary structure of prostaglandin G/H synthase from sheep vesicular gland determined from the complementary DNA sequence. Proc Natl Acad Sci U S A 85:1412–1416

Dinchuck JE, Car BD, Focht RJ, Johnston JJ, Jaffee BD, Covington MB, et al. (1995) Renal abnormalities and an altered inflammatory response in mice lacking cyclooxygenase II. Nature 378:406–409

Distel M, Mueller, C, Bluhmki E, Fries J (1996) Safety of meloxicam: a global analysis of clinical trials. Br J Rheumatol 35[suppl 1]:68–77

Dreser H (1899) Pharmacologisches über Aspirin (Acetylsalicyl-saüre). Pflügers Arch 76:306–318

Eberhart CE, Coffey RJ, Radhika A, Giardiello FM, Ferrenbach S, DuBois RN (1994) Up-regulation of cyclooxygenase 2 gene expression in human colorectal adenomas and adenocarcinomas. Gastroenterology 104:1183–1188

Ehrich E, Schnitzer T, Kivitz A, Weaver A, Wolfe F, Morrison E, Zeng Q, Bolognese J, Seidenberg B et al. (1997) MK-966, a highly selective COX-2 inhibitor, was effective in the treatment of osteoarthritis (OA) of the knee and hip in a 6-week placebo controlled study. Arthritis Rheum 40[suppl 9]:S85

Ehrich EW, Dallob A, De Lepleire I, Van Hecken A, Riendeau D, Yuan W, Porras A, Wittreich J, Seibold JR, De Schepper P, Mehlisch DR, Gertz B (1999) Characterization of rofecoxib as a cyclooxygenase-2 isoform inhibitor and demonstration of analgesia in the dental pain model. Clin Pharmacol Ther 65:336–347

Ferreira SH, Moncada S, Vane JR (1971) Indomethacin and aspirin abolish prostaglandin release from spleen. Nature 231:237–239

Ford-Hutchinson AW (1998) New highly selective COX-2 inhibitors. In: Vane J, Botting J (eds) Selective COX-2 inhibitors. Pharmacology, clinical effects and therapeutic potential. Kluwer Academic, London, pp 117–125

Fries J (1996) Toward an understanding of NSAID-related adverse events: the contribution of longitudinal data. Scand J Rheumatol Suppl 102:3–8

Fu J-Y, Masferrer JL, Seibert K, Raz A, Needleman P (1990) The induction and suppression of prostaglandin $H_2$ synthase (cyclooxygenase) in human monocytes. J Biol Chem 265:16737–16740

Garcia-Rodriguez LA, Jick H (1994) Risk of upper gastrointestinal bleeding and perforation associated with individual non-steroidal anti-inflammatory drugs. Lancet 343:769–772

Garcia-Rodriguez LA, Cattaruzzi C, Troncom MG, Agostinis L (1998) Risk of hospitalization for upper gastrointestinal tract bleeding associated with ketorolac, other nonsteroidal anti-inflammatory drugs, calcium antagonists, and other antihypertensive drugs. Arch Intern Med 158:33–39

Gibb W, Sun M (1996) Localization of prostaglandin H synthase type 2 protein and mRNA in term human fetal membranes and decidua. J Endocrinol 150 497–503

Giuliano F, Warner TD (1999) Ex vivo assay to determine the cyclooxygenase selectivity of non-steroidal anti-inflammatory drugs. Br J Pharmacol 126:1824–1830

Glaser KB (1995) Cyclooxygenase selectivity and NSAIDs: cyclooxygenase-2 selectivity of etodolac (Lodine). Inflammopharmacology 3:335–345

Gustafson-Svärd C, Lilja I, Hallböök O, Sjödahl R (1996) Cyclooxygenase-1 and cyclooxygenase-2 gene expression in human colorectal adenocarcinomas and in azoxymethane induced colonic tumours in rats. Gut 38:79–84

Hamberg M, Svensson J, Samuelsson B (1975) Thromboxanes: a new group of biologically active compounds derived from prostaglandin endoperoxides. Proc Natl Acad Sci U S A 72:2994–2998

Harris RC (1996) The macula densa: recent developments. J Hypertens 14:815–822

Harris RC, McKanna JA, Akai Y, Jacobson HR, Dubois RN, Breyer MD (1994) Cyclooxygenase-2 is associated with the macula densa of rat kidney and increases with salt restriction. J Clin Invest 94:2504–2510

Hawkey C, Kahan A, Steinbruck K, et al. (1998) Gastrointestinal tolerability of meloxicam compared to diclofenac in osteoarthritis patients. Br J Pharmacol 37:937–945

Hemler M, Lands WEM, Smith WL (1976) Purification of the cyclo-oxygenase that forms prostaglandins. Demonstration of the two forms of iron in the holoenzyme. J Biol Chem 251:5575–5579

Henry D, Lim LL-Y, Garcia Rodriguez LA, Perez Gutthann S, Carson JL, Griffin M, et al. (1996) Variability in risk of gastrointestinal complications with individual non-steroidal anti-inflammatory drugs: results of a collaborative meta-analysis. BMJ 312:1563–1566

Hubbard RC, Mehlisch DR, Jasper DR, Nugent MJ, Yu S, Isakson PC (1996) SC-58635, a highly selective inhibitor of COX-2, is an effective analgesic in an acute post-surgical pain model. J Invest Med 44:293A

Isakson P, Zweifel B, Masferrer J, Koboldt C, Seibert K, Hubbard R, et al. (1998) Specific COX-2 inhibitors: from bench to bedside. In: Vane J, Botting J (eds) Selective COX-2 inhibitors. Pharmacology, clinical effects and therapeutic potential. Kluwer Academic, London, pp 1–17

Jackson LM, Wu K, Mahida YR, et al. (1998) COX-1 expression in human gastric mucosa infected with *Helicobacter pylori*: constitutive or induced? Gastroenterology 114:A160

Kargman S, Charleson S, Cartwright M, Frank J, Riendeau D, Mancini J, et al. (1996) Characterization of prostaglandin G/H synthase 1 and 2 in rat, dog, monkey and human gastrointestinal tracts. Gastroenterology 111:445–454

Kawamori T, Rao CV, Seibert K, Reddy BS (1998) Chemopreventive activity of celecoxib, a specific cyclooxygenase-2 inhibitor, against colon carcinogenesis. Cancer Res 58:409–412

Kujubu, DA, Fletcher BS, Varnum BC, Lim RW, Herschman HR (1991) TIS10, a phorbol ester tumor promoter-inducible mRNA from Swiss 3T3 cells, encodes a novel prostaglandin synthase/cyclooxygenase homologue. J Biol Chem 266:12866–12872

Kutchera W, Jones DA, Matsunami N, Groden J, McIntyre TM, Zimmerman GA, et al. (1996) Prostaglandin H synthase 2 is expressed abnormally in human colon cancer: evidence for a transcriptional effect. Proc Natl Acad Sci U S A 93:4816–4820

Laine L, Sloane R, Ferretti M, Cominelli F (1995) A randomised double-blind comparison of placebo, etodolac and naproxen on gastrointestinal injury and prostaglandin production. Gastrointest Endosc 42:428–433

Langenbach R, Morham, SG, Tiano HF, Loftin CD, Ghanayem BI, Chulada PC, et al. (1995) Prostaglandin synthase 1 gene disruption in mice reduces arachidonic acid-induced inflammation and indomethacin-induced gastric ulceration. Cell 83:483–492

Lanza FL (1989) A review of gastric ulcer and gastroduodenal injury in normal volunteers receiving aspirin and other non-steroidal anti-inflammatory drugs. Scand J Gastroenterol Suppl 163:24–31

Lanza FL, Rack MF, Callison DA, Hubbard RC, Yu SS, Talwalker S, et al. (1997a) A pilot endoscopic study of the gastroduodenal effects of SC-58635, a novel COX-2 selective inhibitor. Gastroenterology 112:A194

Lanza F, Simon T, Quan H, Bolognese J, Rack MF, Hoover M, et al. (1997b) Selective inhibition of cyclooxygenase-2 (COX-2) with MK-0966 (250 mg q.i.d.) is associated with less gastroduodenal damage than aspirin (ASA) 650 mg q.i.d. or ibuprofen (IBU) 800 mg t.i.d. Gastroenterology 112:A194

Luk GD (1996) Prevention of gastrointestinal cancer – the potential role of NSAIDs in colorectal cancer. Schweiz Med Wochenschr 126:801–812

Luong C, Miller A, Barnett J, Chow J, Ramesha C, Browner MF (1996) Flexibility of the NSAID binding site in the structure of human cyclooxygenase-2. Nat Struct Biol 3:927–933

MacLagan TJ (1876) The treatment of acute rheumatism by salicin. Lancet 1:342–383

Marcheselli VL, Bazan NG (1996) Sustained induction of prostaglandin endoperoxide synthase-2 by seizures in hippocampus. J Biol Chem 271:24794–24799

Masferrer JL, Zweifel BS, Seibert K, Needleman P (1990) Selective regulation of cellular cyclooxygenase by dexamethasone and endotoxin in mice. J Clin Invest 86:1375–1379

McAdam BF, Catella-Lawson F, Mardini IA, Kapoor JA, Lawson Fitzgerald GA (1999) Systemic biosynthesis of prostacyclin by cyclooxygenase (COX)-2: the human pharmacology of a selective inhibitors of COX-2. Proc Natl Acad Sci U S A 96:272–277

McGeer PL, McGeer EG (1995) The inflammatory response system of brain: implications for therapy of Alzheimer and other neurodegenerative diseases. Brain Res Brain Res Rev 21:195–218

McLaughan J, Seth R, Cole AT (1996) Increased inducible cyclooxygenase associated with treatment failure in ulcerative colitis. Gastroenterology 110:A964

Meade EA, Smith WL, DeWitt DL (1993) Differential inhibition of prostaglandin endoperoxide synthase (cyclooxygenase) isozymes by aspirin and other non-steroidal anti-inflammatory drugs. J Biol Chem 268:6610–6614

Merlie JP, Fagan D, Mudd J, Needleman P (1988) Isolation and characterization of the complementary DNA for sheep seminal vesicle prostaglandin endoperoxide synthase (cyclooxygenase). J Biol Chem 263:3550–3553

Mitchell JA, Akarasereenont P, Thiemermann C, Flower RJ, Vane JR (1993) Selectivity of nonsteroidal anti-inflammatory drugs as inhibitors of constitutive and inducible cyclooxygenase. Proc Natl Acad Sci U S A 90:11693–11697

Mizuno H, Sakamoto C, Matsuda K, Wada K, Uchida T, Noguchi H, Akamatsu T, Kasuga M (1997) Induction of cyclooxygenase 2 in gastric mucosal lesions and its inhibition by the specific antagonist delays healing in mice. Gastroenterology 112:387–397

Moncada S, Gryglewski R, Bunting S, Vane JR (1976) An enzyme isolated from arteries transforms prostaglandin endoperoxides to an unstable substance that inhibits platelet aggregation. Nature 263:663–665

Morham SG, Langenbach R, Loftin CD, Tiano HF, Vouloumanos N, Jenette JC, et al. (1995) Prostaglandin synthase 2 gene disruption causes renal pathology in the mouse. Cell 83:473–482

Nakatsugi S, Fukutake M, Takahashi M, Fukuda K, Isoi T, Taniguchi Y, et al. (1997) Suppression of intestinal polyp development by nimesulide, a selective cyclooxygenase-2 inhibitor, in Min mice. Jpn J Cancer Res 88:1117–1120

O'Banion MK, Sadowski HB, Winn V, Young DA (1991) A serum- and glucocorticoid-regulated 4-kilobase mRNA encodes a cyclooxygenase-related protein. J Biol Chem 266:23261–23267

Oshima M, Dinchuk JE, Kargman SL, Oshima H, Hancock B, Kwong E, et al. (1996) Suppression of intestinal polyposis in $Apc^{D716}$ knockout mice by inhibition of cyclooxygenase 2 (COX-2). Cell 87:803–809

Pairet M, van Ryn J (1998) Tests for cyclooxygenase-1 and -2 inhibition. In: Vane JR, Botting RM (eds) Clinical significance and potential of selective COX-2 inhibitors. William Harvey Press, London, pp 19–30

Palmer MA, Piper PJ, Vane JR (1970) The release of RCS from chopped lung and its antagonism by anti-inflammatory drugs. Br J Pharmacol 40:581P

Pasinetti GM, Aisen PS (1998) Cyclooxygenase-2 expression is increased in frontal cortex of Alzheimer's disease brain. Neuroscience 87:319–324

Patrignani P, Panara MR, Greco A, Fusco O, Natoli C, Iacobelli S, Cipollone F, Ganci A, Creminon C, et al. (1994) Biochemical and pharmacological characterization of the cyclooxgenase activity of human blood prostaglandin endoperoxide synthases. J Pharmacol Exp Ther 271:1705–1712

Patrignani P, Panara MR, Santini G, Sciulli MG, Padovani R, Cipollone F, et al. (1996) Differential inhibition of cyclooxygenase activity of prostaglandin endoperoxide synthase isozymes in vitro and ex vivo in man. Prostaglandins Leukot Essent Fatty Acids 55[suppl 1]:P115

Picot D, Loll PJ, Garavito RM (1994) The X-ray crystal structure of the membrane protein prostaglandin $H_2$ synthase-1. Nature 367:243–249

Piper PJ, Vane JR (1969) The release of prostaglandins during anaphylaxis in guinea-pig isolated lungs. In: Mantegazza P, Horton EW (eds) Prostaglandins, peptides and amines. Academic, London, pp 15–19

Roth GJ, Stanford N, Majerus PW (1975) Acetylation of prostaglandin synthetase by aspirin. Proc Natl Acad Sci U S A 72:3073–3076

Sawdy R, Slater D, Fisk N, Edmonds DK, Bennett P (1997) Use of a cyclooxygenase type-2-selective non-steroidal anti-inflammatory agent to prevent preterm delivery. Lancet 350:265–266

Schassmann A, Peskar BM, Stettlar C, et al. (1998) Effects of inhibition of prostaglandin endoperoxidase synthase-2 in chronic gastro-intestinal ulcer models in rats. Br J Pharmacol 123:795–804

Schneider A, Stahl RAK (1998) Cyclooxygenase-2 (COX-2) and the kidney: current status and potential perspectives. Nephrol Dial Transplant 13:10–12

Schwartz J, Mukhopadhyay S, McBride K, Jones T, Adcock S Sharp P, Hedges K, et al. (1998) Antipyretic activity of a selective cyclooxygenase (COX-2) inhibitor, MK-0966. Clin Pharmacol Therap 63:167

Senna GE, Passalacqua G, Andri G, Dama AR, Albano M, Fregonese L, et al. (1996) Nimesulide in the treatment of patients intolerant of aspirin and other NSAIDs. Drug Saf 14:94–103

Sheng H, Shao J, Kirkland SC, Isakson P, Coffey RJ, Morrow J, et al. (1997) Inhibition of human colon cancer cell growth by selective inhibition of cyclooxygenase-2. J Clin Invest 99:2254–2259

Sirois J, Richards JS (1992) Purification and characterisation of a novel, distinct isoform of prostaglandin endoperoxide synthase induced by human chorionic gonadotropin in granulosa cells of rat preovulatory follicles. J Biol Chem 267:6382–6388

Slater D, Allport V, Bennett P (1998) Changes in the expression of the type-2 but not type-1 cyclooxgenase enzyme in chorion-decidua with the onset of labor. Br J Obstet Gynaecol 105:745–748

Smith JH, Willis AL (1971) Aspirin selectively inhibits prostaglandin production in human platelets. Nature 231:235–237

Spaziani EP, Lantz ME, Benoit RR, O'Brien WF (1996) The induction of cyclooxygenase-2 (COX-2) in intact human amnion tissue by interleukin-4. Prostaglandins 51:215–223

Stewart WF, Kawas C, Corrada M, Metter EJ (1997) Risk of Alzheimer's disease and duration of NSAID use. Neurology 48:626–632

Stone E (1763) An account of the success of the bark of the willow in the cure of agues. Philos Trans R Soc Lond B Biol Sci 53:195–200

Stricker S (1876) Abstract. Dublin J Med Sci 52:395–396

Thun MJ, Namboodiri MM, Heath CWJ (1991) Aspirin use and reduced risk of fatal colon cancer. N Engl J Med 325:1593–1596

Topper JN, Cai J, Falb D, Gimbrone JR (1996) Identification of vascular endothelial genes differentially responsive to fluid mechanical stimuli: cyclooxygenase-2, manganese superoxide dismutase, and endothelial cell nitric oxide synthase are selectively up-regulated by steady laminar shear stress. Proc Natl Acad Sci U S A 93:10417–10422

Vane JR (1971) Inhibition of prostaglandin synthesis as a mechanism of action for aspirin-like drugs. Nat New Biol 231:232–235

Vane JR, Botting RM (1997) Mechanism of action of aspirin-like drugs. Semin Arthritis Rheum 26[suppl 1]:2–10

Vane JR, Bakhle YS, Botting RM (1998) Cyclooxygenases 1 and 2. Annu Rev Pharmacol Toxicol 38:97–120

Warner TD, Giuliano F, Vojnovic I, Bukasa A, Mitchell, JA, Vane JR (1999) Nonsteroid drug selectivities for cyclo-oxygenase-1 rather than cyclo-oxygenase-2 are associated with human gastrointestinal toxicity: a full in vitro analysis. Proc Natl Acad Sci U S A 96:7563–7568

Whittle BJR (1993) Neuronal and endothelium-derived mediators in the modulation of the gastric microcirculation: integrity in the balance. Br J Pharmacol 110:3–17

Whittle BJR, Vane JR (1987) Prostanoids as regulators of gastrointestinal function. In: Johnston LR (ed) Physiology of the gastrointestinal tract, vol 1, 2nd edn. Raven, New York, pp 143–180

Whittle BJR, Higgs GA, Eakins, KE, Moncada S, Vane JR (1980) Selective inhibition of prostaglandin production in inflammatory exudates and gastric mucosa. Nature 284:271–273

Xie W, Chipman, J.G, Robertson DL, Erikson RL, Simmons DL (1991) Expression of a mitogen-responsive gene encoding prostaglandin synthase is regulated by mRNA splicing. Proc Natl Acad Sci U S A 88:2692–2696

Xie W, Robertson DL, Simmons DL (1992) Mitogen-inducible prostaglandin G/H synthase: a new target for nonsteroidal anti-inflammatory drugs. Drug Dev Res 25:249–265

Yokoyama C, Takai T, Tanabe T (1988) Primary structure of sheep prostaglandin endoperoxide synthase deduced from cDNA sequence. FEBS Lett 231:347–351

# 2 Human Group-V Phospholipase-$A_2$ Expression in Pichia pastoris and Its Role in Eicosanoid Generation

L. J. Lefkowitz, H. Shinohara, E. A. Dennis

2.1 Introduction .......................................... 25
2.2 Materials and Methods ................................ 32
2.3 Results ............................................... 39
2.4 Discussion ............................................ 46
References ............................................... 50

## 2.1 Introduction

### 2.1.1 The Importance of Secretory Phospholipase $A_2$s

Ever since the isolation of a non-pancreatic phospholipase $A_2$ ($PLA_2$) from inflammatory exudates of rabbits (Forst et al. 1986) and rats (Chang et al. 1987) and from the synovial fluid of arthritis patients (Stefanski et al. 1986; Kramer et al. 1989; Seilhamer et al. 1989), the group-IIA enzyme has attracted considerable attention as a potential drug target due to its role in the release of arachidonic acid (AA), the precursor of several pro-inflammatory lipids. Many pharmaceutical companies tried to develop inhibitors against this enzyme in the hope of developing therapies for inflammatory diseases. Not surprisingly, much enthusiasm accompanied the discovery of a novel group-V gene (Chen et al. 1994a) that was found in several cell types previously thought to contain only the group-IIA enzyme. These findings have initiated a re-evaluation of earlier studies that had suggested that the group-IIA $PLA_2$ is involved in AA release; these studies did not use methods that could distinguish between the group-IIA and group-V enzymes. After reviewing the discovery of the group-V enzyme and results from cell studies, we describe a novel system that includes the human group-V $PLA_2$ in the methylotrophic yeast *Pichia pastoris*; this system allows the preparation of the enzyme for cellular studies.

### 2.1.2 Discovery of the Group-V PLA2

The cloning and expression of the human group-V $PLA_2$ was first reported in 1994 by Chen et al. (Chen et al. 1994a). A complementary DNA (cDNA) from human brain RNA that encoded a 118 amino-acid peptide with a 20 amino-acid propeptide was obtained. This protein was homologous to other secretory $PLA_2$s ($sPLA_2$s) and had a calculated molecular weight of 13,592 Da, including 12 cysteines that were presumed to form six disulfide bonds. Alignment with other $sPLA_2$s revealed that this putative gene product lacked both the Cys 11–77 disulfide bond and the elapid loop characteristic of group-I enzymes and the six amino-acid C-terminal extension found in group-II enzymes (Figs. 1, 2). Consequently, this novel peptide was classified as a group V enzyme (Tischfield 1997). Northern-blot analysis revealed a 1.2-kb transcript of this gene in heart and lung tissue. To demonstrate that this gene possessed $PLA_2$ activity, the cDNA was expressed in human 293s cells. Elevated levels of $PLA_2$ activity were detected in the culture media using $1$-$^{14}C$ radiolabeled *Escherichia coli* membranes as substrates. The activity was found to be maximal with 10 mM $CaCl_2$ at neutral to alkaline pH, as one would expect for a $sPLA_2$. Substrate specificity was investigated using a thin-layer chromatography assay with radiolabeled phospholipids. These experiments indicated that the group-V $PLA_2$ hydrolyzed in the following order, from most to least strongly hydrolyzed: 1-palmitoyl-2-oleoyl phosphatidylcholine (PC), 1-palmitoyl-2-arachidonoyl PC, 1-palmitoyl-2-arachidonoyl phosphatidylethanolamine (PE), 1-stearoyl-2-arachidonoyl phosphatidylinositol. Later studies reported that the group-IIA and group-V enzymes are very similar, tightly linked and map to human chromosome 1p34-p36.1

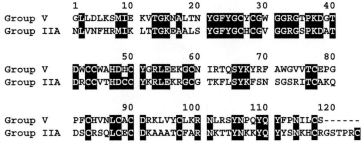

```
               1         10        20        30        40
Group V     GLLDLKSMIE KVTGKNALTN YGFYGCYCGW GGRGTPKDGT
Group IIA   NLVNFHRMIK LTTGKEAALS YGFYGCHCGV GGRGSPKDAT

                        50        60        70        80
Group V     DWCCWAHDHC YGRLEEKGCN IRTQSYKYRF AWGVVTCEPG
Group IIA   DRCCVTHDCC YKRLEKRGCG TKFLSYKFSN SGSRITCAKQ

                        90        100       110       120
Group V     PFCHVNLCAC DRKLVYCLKR NLRSYNPQYQ YFPNILCS------
Group IIA   DSCRSQLCEC DKAAATCFAR NKTTYNKKYQ YYSNKHCRGSTPRC

55 Identical Residues (55/118 * 100 = 46.61%)
```

**Disulfide Bonding Pattern**

| Group IIA | Group V |
|---|---|
| 26-117 | 26-117 |
| 28-44 | 28-44 |
| 43-97 | 43-97 |
| 49-124 | Missing |
| 50-90 | 50-90 |
| 59-83 | 59-83 |
| 77-88 | 77-88 |

**Fig. 1.** Sequence alignment between group-IIA and group-V phospholipase $A_2$ ($PLA_2$). The sequences for mature human group-IIA and group-V $PLA_2$s align perfectly without any insertions or deletions. Identical residues are highlighted in *black*. Based on this alignment, the disulfide-bonding pattern for group-V $PLA_2$ is predicted to be as shown

(Tischfield et al. 1996). These findings suggested that the group-IIA and group-V enzymes arose from a gene-duplication event.

The report by Chen et al. (Chen et al. 1994a) strongly suggested that a new class of $sPLA_2$s had been identified, although additional experiments would be needed to confirm that this enzyme was indeed a novel $PLA_2$. Studies with purified enzyme were required to dismiss the possibility that the $PLA_2$ activity was due to a contaminant. Additionally, assays that are specific for the hydrolysis of the *sn*-2 acyl chain would be

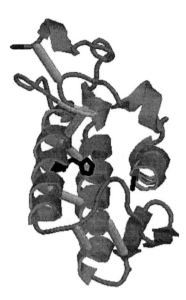

**Fig. 2.** Ribbon diagram of the modeled human group-V phospholipase $A_2$ (PLA$_2$) structure. This view highlights key features of the modeled human group-V PLA$_2$ structure. The six disulfide bonds are represented as cylinders. The calcium-binding loop is colored *white*. This figure was generated using RASMOL (Sayle and Milner-White 1995)

required to demonstrate that this enzyme represented a new class of PLA$_2$s.

Shortly afterward, another paper reported the cloning of group-V PLA$_2$ from rats (Chen et al. 1994b). This enzyme was calcium dependent and had the same substrate specificity as the human group-V enzyme. These findings strengthened the notion that the group-V PLA$_2$s are an important new class of sPLA$_2$s found in mammals.

### 2.1.3 Role of Group-V PLA$_2$ in P388D$_1$ Macrophages

The discovery of the group-V enzyme required a re-evaluation of earlier studies that had attributed PLA$_2$ activity in non-pancreatic tissues to the group-IIA enzyme. Since the group-IIA enzyme is found in inflammatory exudates, many believed that this enzyme was involved in several disease states, including rheumatoid arthritis, septic shock, intestinal neoplasia and cell signaling. Because of the discovery of the group-V $PLA_2$, the $P388D_1$ mouse macrophage-like cell line used by our laboratory was re-examined to determine if it contained the group-V $PLA_2$ (Balboa et al. 1996). Northern-blot analysis of both resting and activated $P388D_1$ cells showed that this cell line contained messenger RNA (mRNA) for group-V $PLA_2$ but not group-IIA $PLA_2$. Similar results were obtained using a reverse-transcriptase polymerase chain reaction (RT-PCR). Additionally, antisense oligonucleotides against the group-V $PLA_2$ decreased the production of $sPLA_2$ and caused a reduction in prostaglandin $E_2$ ($PGE_2$) production (Balboa et al. 1996). These results suggested that the $P388D_1$ macrophages do not contain the group-IIA enzyme as was previously thought; instead, they express the group-V enzyme.

Studies in our laboratory using the murine macrophage-like cell line $P388D_1$ have demonstrated that both the group-IV and the group-V enzymes (Balsinde and Dennis 1996) are involved in the liberation of AA destined to be converted into prostaglandins. Upon stimulation with lipopolysaccharide (LPS)/platelet-activating factor, there is a rapid release of intracellular AA that is attributable to the group-IV $PLA_2$. This intracellular release of AA is required in order to observe group-V $PLA_2$ enzyme on the extracellular leaflet of the plasma membrane. Some of the AA that is released by the group-V enzyme may be taken up by the same cell or neighboring cells and is converted into $PGE_2$ and other prostaglandins via cyclo-oxygenase 2 (COX-2; Balsinde et al. 1998). Using the monoclonal antibody (MAB) clone of the $P388D_1$ cell line, it was found that, during delayed $PGE_2$ production (6–24 h after exposure to LPS alone), induction of both COX-2 and the group-V enzyme is required. In these studies, $PGE_2$ production could be blocked by antisense oligonucleotides against either the group-V enzyme or LY311727, a $sPLA_2$ inhibitor (Shinohara et al. 1999). Direct studies using exogenous human group-V $PLA_2$ in $P388D_1$ macrophage cells have confirmed an important role for this enzyme in COX-2 induction (Balsinde et al. 1999).

### 2.1.4 Studies in Mast and Other Cell Types

Studies were also undertaken to determine whether the group-IIA or the group-V $PLA_2$ is responsible for AA release in MMC-34 mast cells (Reddy et al. 1997). When mast cells are activated by the antigen aggregation of cell-surface, high-affinity immunoglobulin E (IgE) receptors, they degranulate and release a variety of compounds that play an important role in the immune response. In addition to secreting histamine and serotonin, activated mast cells induce the synthesis and release of leukotrienes and $PGD_2$. Previous studies have shown that $PGD_2$ production occurs in two stages: an early phase that takes 10–15 min and a delayed phase that peaks after 4–6 h. Early-phase $PGD_2$ production is attributable to pre-existing COX-1 and requires the action of a $sPLA_2$, while late-phase $PGD_2$ release requires induction of COX-2.

The early-phase $sPLA_2$ was thought to be the group-IIA enzyme, because $PGD_2$ release was inhibited by group-IIA inhibitors and by an antibody against the group-IIA enzyme. Surprisingly, BJ mast cells derived from the bone marrow of mice with a mutation in the Pla2g2a gene (confirmed by Southern-blot analysis) also show early-phase $PGD_2$ synthesis. This finding led to further experiments to see if group-V $PLA_2$ was involved.

Northern-blot analysis and RT-PCR revealed that AKJ mast cells containing the Pla2g2a gene do not express mRNA for the group-IIA enzyme. However, experiments with both AKJ and BJ mast cells showed nearly identical $PGD_2$ induction profiles. Following activation, increased $PLA_2$ activity is also found in the media. Northern-blot analysis and RT-PCR both demonstrate that these cells lines express mRNA for group-V $PLA_2$. The same antisense oligonucleotides against the group-V gene used with macrophages (Balboa et al. 1996) also inhibited $PGD_2$ synthesis and the secretion of $PLA_2$ activity in macrophages. These studies suggest that group-V $PLA_2$ is responsible for early-phase AA mobilization in mast cells and highlight the need to re-evaluate previous studies that implicated group-IIA involvement based on studies using inhibitors and antibodies against the group-IIA enzyme. Growing evidence suggests that antibodies and inhibitors against the group-IIA enzyme also cross-react with the group-V enzyme. Thus, there is a need to develop specific antibodies and inhibitors that can distinguish between these similar enzymes.

Recently, Murakami et al. (Murakami et al. 1998) have carried out studies to understand the roles of cytosolic PLA$_2$ (cPLA$_2$), Ca$^{2+}$-independent PLA$_2$ and sPLA$_2$ (groups IIA, IIC and V, respectively) by overexpressing them in stably transfected human embryonic kidney 293 fibroblasts and Chinese-hamster ovary cells. In these cells, both the recombinant group-IIA and the group-V enzymes are involved in delayed AA release although, when expressed at higher levels, they also participated in immediate AA release. These studies suggest that the group-IIA and group-V PLA$_2$s may serve redundant or overlapping roles. Care should be exercised when extrapolating results obtained with transfected cells to normal cells. It should be noted that some cell types, such as the P388D$_1$ macrophage-like cells and the MMC-34 mast cells, appear to express only the group-V PLA$_2$.

To investigate coupling between various PLA$_2$s and COXs in immediate and delayed prostaglandin-biosynthesis pathways, Murakami et al. (Murakami et al. 1999) expanded their previous studies by transfecting 293 fibroblasts with both COX-1 and COX-2 enzymes, either alone or in combination with various PLA$_2$s. When AA was released by either cPLA$_2$, group-IIA or group-V PLA$_2$, both COX-1 and COX-2 produced PGE$_2$ during the immediate response. In the delayed response, COX-2 was largely responsible for PGE$_2$ production. Interestingly, when cells transfected with group-IIA or group-V PLA$_2$ were co-cultured with non-transfected cells, they increased the production of PGE$_2$ by neighboring cells. Thus, these sPLA$_2$s may play an important physiological role as paracrine amplifiers of signals in order to increase prostaglandin biosynthesis. Since sPLA$_2$s are secreted, function extracellularly and bind to cell surfaces, it makes sense that they are involved in paracrine cell–cell signaling.

## 2.2 Materials and Methods

### 2.2.1 Reagents

Phospholipids were purchased from Avanti Polar Lipids, Inc. (Alabaster, Alabama). *P. pastoris* strains, vectors and enterokinase were from Invitrogen Corporation (Carlsbad, Calif.). Oligonucleotide primers were synthesized by Genosys (Woodlands, Texas). Radiolabeled phospholipids, chromatography resins, restriction enzymes and DNA-modifying enzymes were from Amersham Pharmacia Biotech (Piscataway, New Jersey). All other chemicals used were of at least analytical grade and were obtained from either Fisher or Sigma.

### 2.2.2 Cloning

The pCH10 vector containing the gene for human group-V $PLA_2$ was obtained from Jay A. Tischfield of Indiana University. The group-V gene was released by digestion with *Nhe*I and *Xho*I. To facilitate the transfer of the group-V gene into the pPIC9K *P. pastoris* expression vector, PCR was carried out to introduce a 5' *Xho*I site using the following primers: GGG GGG GGG GCT CGA GAA AAGA and ATG AAA GGC CTC CTC CCA CTG. Digestion of this PCR fragment with *Xho*I and *Eco*RI permitted the incorporation of the group-V gene into the pPIC9 vector (which had also been linearized with *Xho*I and *Eco*RI) to form pPIC9G5. The pPIC9 vector is identical to the pPIC9K vector except that it lacks the kanamycin gene useful for selecting high copy-number transformants; consequently, it lacks an extra *Xho*I site that would have complicated cloning steps (White et al. 1995). To transfer the group-V gene into pPIC9K for expression in *P. pastoris*, pPIC9G5 was digested with *Sac*I and *Sal*I. The *Sac*I/*Sal*I fragment containing the gene for group-V $PLA_2$ was then ligated into the pPIC9K vector (which had been linearized by digestion with *Sac*I and *Sal*I) to form pPIC9KG5. This construct was used to transform the protease-deficient *P. pastoris* strain SMD1168. After induction, abundant $PLA_2$ activity was found in the crude *P. pastoris* media, but attempts to purify the enzyme were unsuccessful.

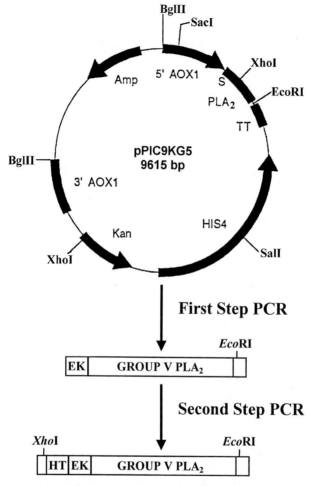

**Fig. 3.** Adding a His tag to the group-V gene using the polymerase chain reaction (PCR). A two-step PCR procedure was used to transfer the group-V gene from the pPIC9KG5 vector and incorporate a His tag that could be removed by treatment with enterokinase to generate native group-V phospholipase $A_2$ without any N-terminal modifications. Primers and PCR conditions are described in Sect. 2.2

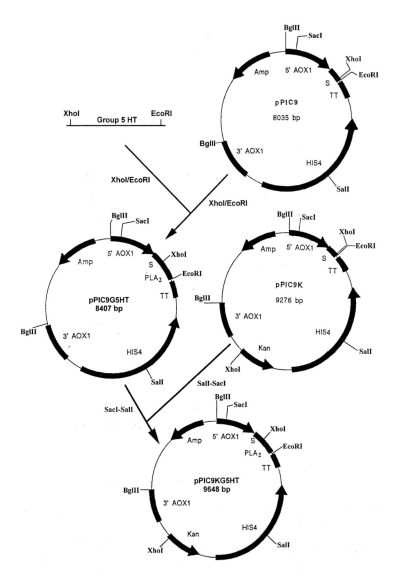

**Fig. 4.** Legend see p. 35

To simplify the purification of human group-V PLA$_2$ from *P. pastoris*, we engineered an N-terminal His tag on the group-V gene; this tag could be removed by treatment with enterokinase. The removable His tag was constructed using a two-step PCR strategy, as shown in Fig. 3. In the first step of the PCR, the 5' portion of the gene was modified to include an enterokinase site using the 48-mer primer (CAC CAC CAC GAC GAT GAC GAT AAA GGC TTG CTG GAC CTA AAA TCA ATG) and a 3' primer that is complementary to a region of the pPIC9K vector located after the stop codon (CGA ATT AAT TCG CGG CCG CCC TAG GG). The PCR reactions contained 50 ng of template DNA, 125 pg of primers, 2.5 mM of each deoxynucleoside triphosphate and 2.5 units of cloned *Pfu* polymerase (Stratagene, San Diego, Calif.) in a 50-µl final volume. Polymerase was added at 95°C to achieve a hot start; 30 cycles of PCR were then carried out, with denaturing at 95°C for 30 s, annealing at 60°C for 1 min and extension for 1 min at 68°C followed by 10 min at 72°C using an MJ Research Minicyler. The product of this first step was then purified and used as template in a second PCR step using the 51-mer 5' primer (GGG GGG TCT CTC GAG AAA AGA CAC CAC CAC CAC CAC CAC GAC GAT GAC GAT) and the same 3' primer and conditions that were used in the first step of the PCR. The product of the second PCR step was a construct containing a 5' *Xho*I site, His tag, enterokinase site, group-V gene and a 3' *Eco*RI site. To transfer this modified construct back into the pPIC9K vector, we followed the procedure that was initially used to transfer the group-V gene into the pPIC9K vector. These steps are shown schemati-

◄

**Fig. 4.** Construction of pPIC9KG5HT. The gene encoding the human group-V phospholipase A$_2$ (*PLA$_2$*) was modified by polymerase chain reaction (PCR) to incorporate a 5' *Xho*I site and a 3' *Eco*RI site, as described in Fig. 3. The group-V gene was digested with *Xho*I and *Eco*RI and subcloned into the *Xho*I/*Eco*RI linearized pPIC9 vector. The resultant vector pPIC9G5HTPLA$_2$ was next digested with *Sac*I/*Sal*I and subcloned into the *Sac*I/*Sal*I linearized pPIC9K vector. This step facilitated incorporation of the PLA$_2$ gene into the pPIC9K vector, which is identical to the pPIC9 vector except that it contains the kanamycin-resistance gene, which is important for the selection of multi-copy transformants in *Pichia pastoris*. The use of pPIC9 as a shuttle vector simplified the cloning steps, because direct digestion of pPIC9K with *Xho*I would have resulted in fragments of nearly identical size (White et al. 1995). The resultant pPIC9KG5HT vector was confirmed by DNA sequencing

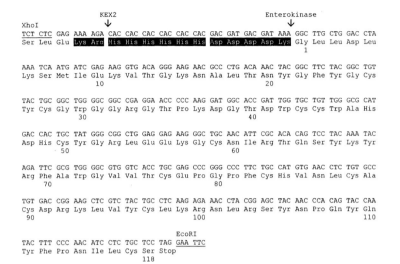

**Fig. 5.** Nucleotide and amino acid sequences of human group-V phospholipase $A_2$ (PLA$_2$). The construct located between the *Xho*I and *Eco*RI restriction sites in the *Pichia pastoris* expression vector pPIC9KG5HT is shown. Restriction sites are *underlined*. Arrows indicate sites of proteolytic cleavage. The locations of the Lys–Arg KEX2 recognition sequence, His tag and (Asp)$_4$–Lys enterokinase recognition site are highlighted in *black*. Removal of the His tag via enterokinase digestion results in the generation of mature group-V PLA$_2$ without any N-terminal modifications

cally in Fig. 4. The sequence of the entire group-V gene is shown in Fig. 5 and was confirmed by automated DNA sequencing (from the region preceding the *Xho*I site to the region following the *Eco*RI site) using an Applied Biosystems 373 Automated DNA Sequencer at the University of California San Diego's Center for Acquired Immune Deficiency Syndrome-Research Molecular-Biology Core.

### 2.2.3 Transformation and Strain Selection

The protease-deficient *P. pastoris* strain SMD1168 was transformed using 10 μg of the *Bgl*II fragment of pPIC9KG5HT that contained the gene of interest, using the same spheroplasting protocol described elsewhere (Balsinde et al. 1999). Colonies containing multiple copies of the group-V $PLA_2$ gene insert were selected by replica plating to yeast extract/peptone/dextrose plates containing 2 g/l of the antibiotic G418. Typically, ten colonies would grow on the G418 plates, and these were screened for $PLA_2$ activity. The strain with the highest activity was chosen for high-level expression using a fermentor.

### 2.2.4 Fermentation and Protein Purification

Human group-V $PLA_2$ was produced using a 5.0 l BioFlow 3000 fermentor (New Brunswick Scientific, Edison, New Jersey) using conditions similar to those used to express the group-IA $PLA_2$ from *Naja naja naja* (Lefkowitz et al., in preparation). Notably, the group-V producing strain had a particularly Mut$^s$ phenotype and grew very slowly in methanol. For this reason, the addition of methanol was ramped more slowly to initiate induction and did not attain the levels required during fermentation of cobra $PLA_2$.

Prior to methanol induction, the glycerol feed was disconnected, and the cells were allowed to deplete any glycerol remaining in the media for 30 min. Induction was initiated by ramping the addition of a 50% methanol solution containing 6 ml/l of PTM4 salts from 1.8 ml/h to 23 ml/h over 30 h. Cells were then switched to a 100% methanol solution containing 12 ml/l of PTM4 salts at a rate of 13 ml/h. Activity was monitored and typically peaked after 48 h of methanol induction. Runs were terminated at this time, and purification was initiated. Recombinant group-V $PLA_2$ was purified as described elsewhere (Balsinde et al. 1999). Protein concentrations were determined with the BioRad microassay using bovine serum albumin as the standard.

### 2.2.5 Enzyme Assay

$PLA_2$ activity was measured using the Dole assay, as described previously (Conde-Frieboes et al. 1996). Briefly, substrate was prepared by drying phospholipid stocks in chloroform under a stream of nitrogen. Dried lipid was resuspended in Triton X-100 and diluted with buffer to a final concentration of 100 µm phospholipid, 400 µM Triton X-100,

25 mM tris-(hydroxymethyl)-aminomethane (Tris)-Cl, pH 8.5 and 10 mM $CaCl_2$. Sufficient radiolabeled 1-palmitoyl-2-[1-$^{14}$C]palmitoyl PC was included to provide 100,000 cpm per assay. Reactions were initiated by adding 400 µl of substrate to 100 µl of enzyme and were allowed to proceed for 30 min at 40°C, with agitation. To quench the reaction, 2.5 ml of Dole reagent [2-propanol:heptane:1 N $H_2SO_4$, 20:5:1 (v/v/v)] were added, followed by approximately 0.1 g silica gel, 1.5 ml heptane and 1.5 ml deionized water. After vortexing and phase separation, 1 ml of the upper heptane layer was removed and passed through a plugged Pasteur pipet containing approximately 0.1 g silica gel. To collect any remaining radiolabeled fatty acid, the column was washed with 1 ml of ether. Prior to counting on a Packard TR1600 Liquid Scintillation Analyzer, 5 ml scintillation fluid was added to each sample.

### 2.2.6 Western Blotting

Proteins were separated on 18% Tris–glycine gels (Novex) and were transferred to Immobilon-P membranes (Millipore). Non-specific binding was blocked by incubating membranes in 5% nonfat milk in phosphate-buffered saline (Sigma) for 1 h. An antibody against human group-V $PLA_2$ that was expressed in *E. coli* (Chen and Dennis 1998) was obtained from Dr. Yijun Chen and was diluted in blocking buffer (1:500) and incubated with the membrane for 1 h. Protein A conjugated to horseradish peroxidase was used as the secondary antibody and facilitated detection by enhanced chemiluminescence (Amersham Pharmacia Biotech, Piscataway, N.J.).

To detect proteins containing a His tag, Western blots were carried out using a MAB against a His tag that was obtained from Qiagen (Valencia, Calif.) and used according to the manufacturer's recommendations. It was found to be essential to carry out all washes in Tris-Cl buffered saline [TBS; 10 mM Tris-Cl (pH 7.5), 150 mM NaCl] containing 0.05% (v/v) Tween-20 and 0.2% (v/v) TX-100. Blocking buffer consisted of 1% (w/v) casein in TBS. Goat anti-mouse IgG conjugated to horseradish peroxidase (Jackson Immunoresearch) was used as the secondary antibody and facilitated detection by enhanced chemiluminescence (Amersham Pharmacia Biotech, Piscataway, N. J.).

## 2.2.7 Molecular Modeling

Based on the extraordinary degree of similarity between the group-IIA and the group-V $PLA_2$ sequences, a model of the group-V enzyme was constructed using Insight II software (Molecular Simulations Inc., San Diego, Calif.). The uninhibited group-IIA structure reported by Scott et al. (Scott et al. 1991) was used as a template, and 63 side chain substitutions were made to transform the group-IIA enzyme into the group-V enzyme. The quality of the group-V model was assessed using Pro-Check and was found to be similar to that of the starting structure. Some of the key features of the modeled group-V structure are shown in Fig. 2.

## 2.3 Results

### 2.3.1 Molecular-Modeling Studies

One of the first things to do prior to attempting to express any protein in *P. pastoris* is to scan the sequence for any N-linked glycosylation consensus sequences or internal KEX2 sites. The human group-V $PLA_2$ sequence did not contain any glycosylation sites, but there was at least one internal KEX2 site. KEX2 is a protease that cleaves the α-factor secretion signal in yeast after dibasic Lys–Arg sequences. Analysis of the group-V sequence reveals that Arg-91/Lys-92 might be cleaved by KEX2, and Lys-99/Arg-100 certainly should be cleaved, provided these residues are not made inaccessible due to folding. To predict whether the charged groups from both of these dibasic regions are surface exposed, we attempted to model the structure of the human group-V $PLA_2$.

Based on sequence alignments with other sPLA2s, the human group-V enzyme was placed in its own group. This 118 amino acid protein has a calculated molecular weight of 13.6 kDa but lacks the Cys-11/Cys-77 disulfide bond and elapid loop that are characteristic of the group-I enzymes. The group-V enzyme also lacks the C-terminal extension found in group-II enzymes and the Cys-50 to Cys-124 disulfide bond. What is not apparent in a previously published sequence alignment of the human group-IIA and the human group-V enzymes (Tischfield 1997) is that both of these enzymes can be aligned perfectly without the use of any sophisticated computer programs. As shown in Fig. 1, these enzymes simply align without any insertions or deletions, and 47% of their residues are identical. The only difference is that the group-V enzyme lacks the C-terminal extension of the group-IIA enzyme and, therefore, is also missing the Cys-50 to Cys-124 disulfide bond. This degree of similarity is impressive.

Conveniently, the structure of the group-IIA enzyme has been solved by X-ray crystallography in three different crystal forms (Scott et al. 1991; Wery et al. 1991), including one that contains a bound transition-state analog (Scott et al. 1991). As a result of the high degree of identity between the group-IIA and group-V enzymes and the availability of coordinates for the group-IIA structure, it was possible to create a modeled structure of the group-V enzyme, as shown in Fig. 2. In this modeled structure, the amide bonds of Arg-91/Lys-92 and Lys-99/Arg-100 were not surface exposed, suggesting that internal KEX2 cleavage should not pose a problem during expression in *P. pastoris*.

### 2.3.2 Cloning

The gene encoding human group-V $PLA_2$ was modified to include an N-terminal His tag that could be removed via digestion with enterokinase to yield native enzyme without any N-terminal modifications. Figure 3 describes the two-step PCR strategy used to make this construct. In the first step, the enterokinase site was created. The second PCR step completed the His tag and created a *Xho*I site that could be used for cloning into the pPIC9 vector. The primers used to accomplish these modifications are described in the methods section. The nucleotide and amino acid sequences for the human group-V $PLA_2$ gene containing a His tag are shown in Fig. 5. This design facilitated cloning into the pPIC9 *P. pastoris* expression vector at the *Xho*I and *Eco*RI restriction sites. The six histidines that comprise the His tag are highlighted in black. Arrows indicate sites of proteolytic cleavage. The KEX2 protease cleaves after Lys–Arg in the α-factor secretion signal, while enterokinase cleaves after Asp–Asp–Asp–Asp–Lys. Following treatment with enterokinase, native group-V $PLA_2$ is obtained with an intact N-terminus and without any additional amino acids.

The steps required to transfer the group-V gene containing the His tag into the pPIK9K expression vector are shown in Fig. 4. The PCR fragment containing group-V PLA$_2$ was first digested with *Xho*I and *Eco*RI to facilitate ligation into the pPIC9 vector (which had been linearized with *Xho*I and *Eco*RI) to create the pPIC9G5HT vector. The pPIC9 vector is identical to the pPIC9K vector except that it lacks the kanamycin-resistance gene, which is important for selecting high copy-number transformants in *P. pastoris*. Since the pPIC9K vector contains two *Xho*I sites, pPIC9 was used as a shuttle vector to simplify cloning. The group-V containing *Sac*I/*Sal*I fragment of pPIC9G5HT was ligated to the 6300-bp *Sac*I/*Sal*I fragment of pPIC9K to form pPIC9KG5HT. Prior to transformation into yeast, the group-V PLA$_2$ construct was confirmed by DNA sequencing.

### 2.3.3 Expression

Human group-V PLA$_2$ was expressed in the methylotrophic yeast *P. pastoris* using a fermentation process similar to that described previously (Chen et al. 1997). The phenotype of the *P. pastoris* strain was particularly Mut$^s$. Mut$^s$ strains grow slowly in methanol, because the gene of interest has disrupted the AOX1 gene, which encodes alcohol oxidase and enables *P. pastoris* cells to utilize methanol as both an energy and carbon source. These strains must then rely on their AOX2 gene, which is transcribed with lower efficiency; thus, cells tend to grow more slowly when utilizing methanol as their sole carbon source. During fermentation, methanol was fed more slowly to these cells and did not exceed 13 ml/h of 100% methanol containing 12 ml/l PTM4 salts.

### 2.3.4 Purification

Purification was accomplished using metal–chelate affinity chromatography and ion-exchange chromatography. Following fermentation, cells were pelleted using a low-speed centrifugation step, and the crude media was stored frozen. To initiate purification, the crude media was thawed and centrifuged at a higher speed to remove any remaining cells and proteins that had aggregated during freezing. This step reduced the

**Table 1.** Purification of human group-V phospholipase $A_2$. Enzyme activity was determined at 40°C using a modified Dole assay containing 100 μM dipalmitoylphosphatidylcholine, 400 μM TX-100, 10 mM $CaCl_2$, 25 mM tris-(hydroxymethyl)-aminomethane (pH 8.5) and 100 mM KCl. Protein concentrations were determined with the BioRad assay using bovine serum albumin as the standard

| Procedure | Volume (ml) | Total activity (units[a]) | Total protein (mg) | Specific activity (units/mg) | Yield (%) |
|---|---|---|---|---|---|
| Crude media | 1500 | 1323 | 685.5 | 1.93 | 100 |
| Nickel column | 254 | 232 | 8.64 | 26.85 | 17 |
| DEAE column | 3 | 16.2 | 0.06 | 270 | 1 |

*DEAE*, diethylaminoethanol.
[a] One unit is the amount of enzyme required to hydrolyze 1 nmol of substrate per minute.

viscosity of the crude media, enabling it to be applied directly to a gravity flow column.

The first column utilized was a nickel affinity column that binds proteins containing a His tag. Initial attempts at purification resulted in the binding of only small quantities of the group-V $PLA_2$ to the column; however, activity measurements and Western-blot analysis using an antibody against the His tag indicated that similar levels of the group-V enzyme were present in both the crude media and the nickel column eluant. These findings suggested that binding to the column was impaired. One possibility was that the His tag might interact strongly with another part of the group-V enzyme and, therefore, might not be accessible to the nickel resin. To test this possibility, the column was run under strongly denaturing conditions with 6 M guanidine hydrochloride (GdnHCl). This resulted in excellent column binding and high yields. N-terminal sequencing demonstrated that this material was the group-V $PLA_2$. Unfortunately, the enzyme was inactive after denaturation. Since the objective was to obtain properly folded group-V enzyme without any N-terminal modifications, we attempted to optimize conditions under which the enzyme would bind to the nickel resin. It was found that reasonable amounts of enzyme would bind to the column in the presence of 500 mM GdnHCl. On dialysis, the enzyme appeared to regain full

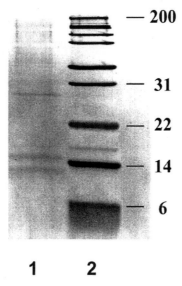

**Fig. 6.** Sodium dodecyl sulfate polyacrylamide-gel electrophoresis of human group-V phospholipase $A_2$ ($PLA_2$). Samples were loaded on 18% tris-(hydroxymethyl)-aminomethane–glycine gels and silver stained. *Lane 1* shows group-V $PLA_2$ following purification via affinity and ion-exchange chromatography. The major band at 13.6 kDa is native group-V $PLA_2$. The 15-kDa band reveals that digestion with enterokinase was not complete. Molecular-weight standards are shown in *lane 2*

enzymatic activity, indicating that the mildly denaturing conditions probably contributed to a more flexible conformation of the N-terminal His tag without causing any major, irreversible changes to the enzyme. Thus, metal–chelate affinity chromatography was carried out under these mildly denaturing conditions to maximize the amount of human group-V $PLA_2$ that could be purified from the crude *P. pastoris* media.

As predicted from the modeled structure of the group-V $PLA_2$, enzyme was expressed in *P. pastoris* without being cleaved internally by KEX2. Enzyme production was routinely checked by Western blots using an antibody against the His tag. These experiments did not reveal any truncation products, confirming that these residues are likely inaccessible. We do not know of any other examples of proteins expressed in

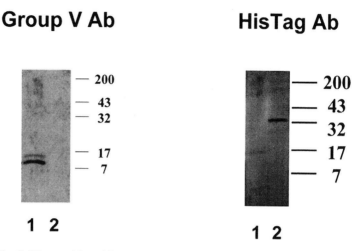

**Fig. 7.** Western blot of human group-V phospholipase $A_2$ ($PLA_2$). Approximately 100 ng of human group-V $PLA_2$ that had been purified on a diethylaminoethanol column and treated with enterokinase (*lane 1*) or human lyso-$PLA_2$ (Wang et al. 1997) that contained a His tag as a control (*lane 2*) were run on 18% tris-(hydroxymethyl)-aminomethane–glycine gels under denaturing conditions and probed using antibodies against either human group-V $PLA_2$ (*left*) or His tag (*right*). Molecular-weight standards are indicated on the right

*P. pastoris* and containing internal KEX2 sites, but it is likely that other such enzymes exist. Provided that the KEX2 protease is unable to gain access to its substrate, these internal dibasic residues should not pose problems. In these instances, it can be invaluable to have either a structure or a modeled structure of the protein of interest to assess whether or not the expression strategy will be successful.

Ion-exchange chromatography was used to further purify the group-V enzyme using a diethylaminoethanol (DEAE) column, as summarized in Table 1. Enzyme bound under low-salt conditions and could be eluted using a linear gradient of 0–1.0 M NaCl. Fractions containing $PLA_2$ activity were pooled and treated with enterokinase to remove the His tag. Removal of the His tag resulted in an approximately sixfold increase in activity using dipalmitoyl phosphatidylcholine (DPPC)-mixed micelles. These results suggest that the N-terminus on the human group-V $PLA_2$ is important for full enzymatic activity.

Purity was assessed using sodium dodecyl sulfate polyacrylamide-gel electrophoresis (SDS-PAGE) with silver staining and Western-blot analysis. Figure 6 shows a SDS-PAGE gel of the group-V $PLA_2$ purified through the DEAE column. The enzyme is purified to near homogeneity, although the presence of a 15-kDa band indicates that the enterokinase reaction did not go to completion. To determine the ratio of cleaved to uncleaved group-V $PLA_2$, a Western blot was probed using an antibody against the group-V enzyme obtained from Dr. Yijun Chen. As shown in Fig. 7, the Western blot reveals that approximately 90% of the His tag has been removed. It is unlikely that the His tag interferes with antibody recognition, because the proteins were separated under denaturing conditions. Thus, the final preparation of the group-V $PLA_2$ is predominantly recombinant enzyme with the correct N-terminus.

### 2.3.5 Activity Measurements

The specific activity of the human group-V $PLA_2$ was found to be 270 nmol/min/mg using mixed micelles containing TX-100 and DPPC. Even after 5 months of storage at 4°C, the group-V enzyme expressed in *P. pastoris* showed only minor loss of activity, demonstrating excellent stability (Table 2). To test for phosphatidyl glycerol (PG) activation, 5 mole percent of 1-palmitoyl-2-oleylphosphatidyl glycerol (POPG) was included in these assays. The presence of POPG led to a 4.7-fold

**Table 2.** Activity of human group-V phosphoplipase $A_2$ ($PLA_2$). The specific activity of human group-V $PLA_2$ in mixed micelles containing 400 μM dipalmitoylphosphatidylcholine (DPPC) and 1.6 mM TX-100 was measured using the Dole assay. To test for activation by prostaglandin, 5 mole percent of 1-palmitoyl-2-oleylphosphatidyl glycerol (POPG) was included. To test for activation by anionic surfaces, 5 mole percent of cholic acid was included. Values are reported as the mean ± standard deviation for duplicate measurements

| Substrate | Detergent | Specific activity (nmol/min/mg) |
|---|---|---|
| DPPC | TX-100 | 230±10 |
| DPPC+POPG | TX-100 | 1070±30 |
| DPPC | TX-100+cholic acid | 510±50 |

increase in the hydrolysis of DPPC. To determine whether this activation might be due to surface charge, 5 mole percent of the anionic detergent cholic acid was included in a separate experiment. The presence of cholic acid led to a 2.2-fold increase in the rate of hydrolysis of DPPC. Thus, approximately half of the activation achieved by PG can be accounted for by a non-specific surface charge.

### 2.3.6 Use of Recombinant Human Group-V PLA₂ in Cell Studies

The human group-V $PLA_2$ produced in *P. pastoris* was recently utilized by our laboratory (Balsinde et al. 1999) to investigate the effect of exogenous group-V $PLA_2$ on COX-2 expression in $P388D_1$ macrophages (MAB clone). Previous studies showed that prolonged exposure to LPS is mediated by COX-2 (Shinohara et al. 1999). Inhibition of group-V $PLA_2$ by LY311727 or antisense oligonucleotides specific for group-V $PLA_2$ were found to cause a dose-dependent decrease in COX-2 expression. In experiments where exogenous group-V $PLA_2$ was added, AA release and COX-2 expression both increased. These effects were not observed if the group-V enzyme was first inhibited with bromophenacyl bromide, implying that active enzyme is required to induce COX-2 expression. These findings are consistent with a model whereby the release of AA by the group-V $PLA_2$ is responsible for delayed prostaglandin production by regulating COX-2 expression. These studies also provide evidence that human group-V $PLA_2$ expressed in *P. pastoris* is biologically active.

## 2.4 Discussion

### 2.4.1 Similarity of the Group-IIA and the Group-V PLA₂s

The classification of group-V $PLA_2$ was based on sequence alignments with other $sPLA_2$s. These comparisons revealed that the group-V enzyme contains 12 cysteines but lacks the Cys-11 to Cys-77 disulfide bond found in the group-I enzymes. The group-V enzyme also lacks the C-terminal extension found in the group-II enzymes. For these reasons, the group-V enzyme was placed in its own group. In the past, it has been recognized that all sPLA$_2$s are characterized by their low molecular weight, millimolar calcium requirement and high percentage of disulfide bonds. Consequently, the disulfide-bonding patterns of these enzymes have served as the defining feature for classifying sPLA$_2$s.

Initial comparison of the group-V PLA2 sequence to that of other sPLA2s reveals a high degree of homology, as one would expect and others (Tischfield 1997) have reported. However, closer examination reveals that the group-IIA and group-V enzymes can be aligned perfectly without the use of any sophisticated computer programs. These enzymes align perfectly without any insertions or deletions, and 47% of their residues are identical. Since the group-V enzyme contains only 118 amino acids, compared with 124 for the group-IIA enzyme, the group-V enzyme lacks both the C-terminal extension and the Cys-50 to Cys-124 disulfide bond of the group-IIA enzyme. A consequence of the slightly different primary structures of these enzymes is that the calculated isoelectric point of the group-IIA enzyme is much more basic than that of the group-V enzyme. Preliminary characterization experiments seem to show that these enzymes are both secreted and calcium dependent and, at least in fibroblasts, they appear to carry out overlapping functions (Murakami et al. 1998).

### 2.4.2 Strategies for Group-V PLA2 Expression

Thus far, two groups have reported the expression of human group-V PLA$_2$ in *E. coli*. The report by Han et al. (Han et al. 1998) utilized a synthetic group-V gene containing codons optimized for bacterial expression. This gene was then subcloned into the pET21a vector for protein expression using the β-galactosidase promoter. After induction with isopropyl β-D-thiogalactopyranoside, the cells were collected by centrifugation, resuspended in buffer and lysed (by sonication) to release inclusion bodies containing aggregated group-V PLA$_2$. Gel-filtration chromatography was used to purify the enzyme prior to refolding. After refolding, the enzyme was further purified using ion-exchange chromatography. Initial characterization experiments using non-natural polymerized liposome substrates revealed that the group-V enzyme hydrolyzed PC substrates about as well as it hydrolyzed PG and phosphatidic acid substrates but showed lower specificity for PE and phosphatidylserine substrates. In this system, group-V $PLA_2$ hydrolyzed PC more efficiently than the group-IIA $PLA_2$. The group-V enzyme was also found to have a higher binding affinity for PC bilayers. Thermal-inactivation studies indicated that the group-V enzyme is less stable than other $sPLA_2s$. The authors suggested that the group-V $PLA_2$ may be better suited to liberating AA from the outer leaflet of PC-rich cell membranes than the group-IIA enzyme and that its low stability may be a form of regulation.

Han et al. (Han et al. 1998) designed a gene construct for the group-V $PLA_2$; this construct incorporates an initiator methionine residue at the N-terminus. There is a substantial body of evidence describing the detrimental effects of N-terminal modifications to $sPLA_2s$ (Scott and Sigler 1994). Although it was assumed that the methionine was cleaved, an experimental demonstration was not provided. Additionally, the authors noted significant problems with protein refolding, citing low yields and problems with precipitation. Lastly, the use of unnatural substrates made it impossible to directly compare the group-V enzyme to other $sPLA_2s$.

Our laboratory also reported the expression of the human group-V $PLA_2$ in *E. coli* (Chen and Dennis 1998). The group-V gene was obtained from human heart mRNA by RT-PCR and subcloned into the pET28a vector. This construct utilized a N-terminal His tag that could be removed by digestion with thrombin, although three amino acids from the thrombin cleavage site and an unused initiator methionine remain on the N-terminus. As in the above case, it is possible that such N-terminal modifications may alter the enzymatic properties of the group-V enzyme, as has been reported with other $sPLA_2s$ (Scott and Sigler 1994). However, the His tag enabled efficient single-column purification of the enzyme using Ni–nitrilotriacetic acid resin. Refolding was carried out prior to removal of the His tag by digestion with biotinylated thrombin. The thrombin could then be removed with streptavidin-linked agarose resin. Purification to homogeneity was assessed by SDS-PAGE.

Characterization experiments using natural phospholipid substrates revealed a calcium dependence and an optimum pH of 8.5. 1-palmitoyl-2-linoleoyl PE (PLPE) vesicles were hydrolyzed more efficiently than 1-palmitoyl-2-linoleoyl PC (PLPC) vesicles; however, when both PLPE and PLPC were present in equal amounts, PLPC was hydrolyzed slightly more efficiently than PLPE, suggesting that the group-V enzyme has little preference for the PC head group versus the PE head group. Experiments to determine acyl-chain specificity for substrates containing a PC head group found that specificity occurred in the order: linoleoyl palmitoyl arachidonoyl. Importantly, this study reported the activity of purified group-V PLA$_2$ on natural substrates. The group-V enzyme was also shown to be inhibited by 3-(3-actamide-1-benzyl-2-ethylindolyl-5-oxy)propane phosphonic acid (LY311727), a potent inhibitor of group-IIA PLA$_2$. IC$_{50}$ values were 23 nM for group-IIA and 36 nM for group-V PLA$_2$, although the experimental conditions for each enzyme were not the same.

Previously, we showed that the *P. pastoris* expression system is useful for expressing cobra venom PLA$_2$ (Lefkowitz et al., in preparation). Advantages of this system include the correct folding of highly disulfide-bonded proteins, protein secretion of expressed proteins directly into the media, endotoxin-free protein production and inexpensive growth media. Since the group-V enzyme is similar to other sPLA$_2$s, our laboratory first attempted to express this enzyme in *P. pastoris*. Initial results indicated that PLA$_2$ activity was being secreted, but attempts to purify the enzyme were not successful. To circumvent this problem, a N-terminal His tag that could be removed via treatment with enterokinase to yield native enzyme without any N-terminal modifications was designed. The decision to produce group-V PLA$_2$ without any N-terminal modifications was based on numerous previous studies that described the importance of the N-terminus for proper enzyme action (Scott and Sigler 1994). Thus, the *P. pastoris* expression system reported here has distinct advantages for the preparation of the properly folded enzyme (which should be identical to the mature group-V enzyme).

These experiments represent the first example of expression of the group-V PLA$_2$ without any refolding steps or N-terminal modifications. The incorporation of a His tag allowed us to monitor protein expression by Western blotting; however, the His tag did not bind tightly to the affinity column. This was presumably due to a strong interaction with a nearby region of the group-V enzyme or possible interactions with another protein. In any event, mildly denaturing conditions permitted sufficient group-V PLA$_2$ to be purified for initial characterization experiments. These findings highlight some of the difficulties encountered when trying to express sPLA$_2$s as fusion proteins. It is important to express these enzymes without any N-terminal modifications, but the rigid N-terminal helix can make it difficult for processing enzymes to cleave between the enzyme and the fusion protein.

**Acknowledgements.** This work was supported by a grant from the National Institutes of Health (NIH), GM 20501. L.J. Lefkowitz was the recipient of NIH pre-doctoral training grant 5 T32 DK07202. The authors wish to acknowledge the kind gift of a New Brunswick BioFlow 3000 Fermentor from Central Soya, Inc.

## References

Balboa MA, Balsinde J, Winstead MV, Tischfield JA, Dennis EA (1996) Novel group V phospholipase A2 involved in arachidonic acid mobilization in murine P388D1 macrophages. J Biol Chem 271:32381–32384

Balsinde J, Dennis EA (1996) Distinct roles in signal transduction for each of the phospholipase A2 enzymes present in P388D1 macrophages. J Biol Chem 271:6758–6765

Balsinde J, Balboa MA, Dennis EA (1998) Functional coupling between secretory phospholipase $A_2$ and cyclooxygenase-2 and its regulation by cytosolic group IV phospholipase $A_2$. Proc Natl Acad Sci U S A 95:7951–7956

Balsinde J, Shinohara H, Lefkowitz LJ, Johnson CA, Balboa MA, Dennis EA (1999) Group-V phospholipase $A_2$ dependent induction of cyclooxygenase-2 in macrophages. J Biol Chem 274:25967–25970

Chang HW, Kudo I, Tomita M, Inoue K (1987) Purification and characterization of extracellular phospholipase $A_2$ from peritoneal cavity of caseinate-treated rat. J Biochem (Tokyo) 102:147–154

Chen J, Engle SJ, Seilhamer JJ, Tischfield JA (1994a) Cloning and recombinant expression of a novel human low molecular weight Ca(2+)-dependent phospholipase $A_2$. J Biol Chem 269:2365–2368

Chen J, Engle SJ, Seilhamer JJ, Tischfield JA (1994b) Cloning, expression and partial characterization of a novel rat phospholipase $A_2$. Biochim Biophys Acta 1215:115–120

Chen Y, Dennis EA (1998) Expression and characterization of human group V phospholipase $A_2$. Biochim Biophys Acta 1394:57–64

Chen Y, Cino J, Hart G, Freedman D, White CE, Komives EA (1997) High protein expression in fermentation of recombinant *Pichia pastoris* by a fed-batch process. Process Biochem 32:107–111

Conde-Frieboes K, Reynolds LJ, Lio Y, Hale M, Wasserman HH, Dennis EA (1996) Activated ketones as inhibitors of intracellular $Ca^{2+}$-dependent and $Ca^{2+}$-independent phospholipase $A_2$. J Am Chem Soc 118:5519–5525

Forst S, Weiss J, Elsbach P, Maraganore JM, Reardon I, Heinrikson RL (1986) Structural and functional properties of a phospholipase $A_2$ purified from an inflammatory exudate. Biochemistry 25:8381–8385

Han SK, Yoon ET, Cho WH (1998) Bacterial expression and characterization of human secretory class V phospholipase A2. Biochem J 331:353–357

Kramer RM, Hession C, Johansen B, Hayes G, McGray P, Chow EP, Tizard R, Pepinsky RB (1989) Structure and properties of a human non-pancreatic phospholipase $A_2$. J Biol Chem 264:5768–5775

Murakami M, Shimbara S, Kambe T, Kuwata H, Winstead MV, Tischfield JA, Kudo I (1998) The functions of five distinct mammalian phospholipase $A_2$s in regulating arachidonic acid release – type IIA and type V secretory phospholipase $A_2$s are functionally redundant and act in concert with cytosolic phospholipase $A_2$. J Biol Chem 273:14411–14423

Murakami M, Kambe T, Shimbara S, Kudo I (1999) Functional coupling between various phospholipase $A_2$s and cyclooxygenases in immediate and delayed prostanoid biosynthetic pathways. J Biol Chem 274:3103–3115

Reddy ST, Winstead MV, Tischfield JA, Herschman HR (1997) Analysis of the secretory phospholipase $A_2$ that mediates prostaglandin production in mast cells. J Biol Chem 272:13591–13596

Sayle RA, Milner-White EJ (1995) RASMOL: biomolecular graphics for all. Trends Biochem Sci 20:374–376

Scott DL, Sigler PB (1994) Structure and catalytic mechanism of secretory phospholipases $A_2$. Adv Protein Chem 45:53–88

Scott DL, White SP, Browning JL, Rosa JJ, Gelb MH, Sigler PB (1991) Structures of free and inhibited human secretory phospholipase $A_2$ from inflammatory exudate. Science 254:1007–1010

Seilhamer JJ, Pruzanski W, Vadas P, Plant S, Miller JA, Kloss J, Johnson LK (1989) Cloning and recombinant expression of phospholipase $A_2$ present in rheumatoid arthritic synovial fluid. J Biol Chem 264:5335–5338

Shinohara H, Balboa MA, Johnson CA, Balsinde J, Dennis EA (1999) Regulation of delayed prostaglandin production in activated P388D1 macrophages by group-IV cytosolic and group-V secretory phospholipase $A_2$. J Biol Chem 274:12263–12268

Stefanski E, Pruzanski W, Sternby B, Vadas P (1986) Purification of a soluble phospholipase $A_2$ from synovial fluid in rheumatoid arthritis. J Biochem (Tokyo) 100:1297–1303

Tischfield JA (1997) A thematic series on phospholipases. 5. A reassessment of the low molecular weight phospholipase $A_2$ gene family in mammals. J Biol Chem 272:17247–17250

Tischfield JA, Xia YR, Shih DM, Klisak I, Chen J, Engle SJ, Siakotos AN, Winstead MV, Seilhamer JJ, Allamand V, Gyapay G, Lusis AJ (1996) Low-molecular-weight, calcium-dependent phospholipase $A_2$ genes are linked and map to homologous chromosome regions in mouse and human. Genomics 32:328–333

Wang AJ, Deems RA, Dennis EA (1997) Cloning, expression, and catalytic mechanism of murine lysophospholipase I. J Biol Chem 272:12723–12729

Wery JP, Schevitz RW, Clawson DK, Bobbitt JL, Dow ER, Gamboa G, Goodson TJ, Hermann RB, Kramer RM, McClure DB (1991) Structure of recombinant human rheumatoid arthritic synovial fluid phospholipase $A_2$ at 2.2 A resolution. Nature 352:79–82

White CE, Hunter MJ, Meininger DP, White LR, Komives EA (1995) Large-scale expression, purification and characterization of small fragments of thrombomodulin: the roles of the sixth domain and of methionine 388. Protein Eng 8:1177–1187

# 3 Fatty-Acid Substrate Interactions with Cyclo-oxygenases

W. L. Smith, C. J. Rieke, E. D. Thuresson, A. M. Mulichak, and R. M. Garavito

| | | |
|---|---|---|
| 3.1 | Introduction | 53 |
| 3.2 | COX and Peroxidase Catalysis | 54 |
| 3.3 | COX Substrates | 56 |
| 3.4 | Native PGHS-1 | 58 |
| 3.5 | Mutant PGHS-1 | 59 |
| 3.6 | Arachidonate Binding to PGHS-1 Versus PGHS-2 | 59 |
| 3.7 | Future Studies | 60 |
| References | | 60 |

## 3.1 Introduction

Prostaglandin endoperoxide H synthases 1 and -2 (PGHS-1 and -2) convert arachidonic acid and $O_2$ (along with two reducing equivalents) to $PGH_2$ – the committing step in the formation of prostanoids (Smith and DeWitt 1996; Smith et al. 1996). PGHS-1 is often referred to as the constitutive enzyme, whereas PGHS-2 is known as the inducible isoform. They differ from one another mainly with respect to their temporal patterns of expression. The reason for the existence of the two PGHS isozymes is still unknown. One possibility is that PGHS-2 is induced and then functions at relatively low fatty-acid substrate and hydroperoxide-activator concentrations to generate prostanoid products during early stages of cell replication or differentiation, whereas PGHS-1 forms products that are involved in "housekeeping" functions when circulating hormones act on cells acutely to cause the release of higher concentrations of arachidonate (Capdevila et al. 1995; Kulmacz and Wang 1995; Kulmacz 1998; So et al. 1998). Structurally (Picot et al. 1994; Kurumbail et al. 1996; Luong et al. 1996) and kinetically (Barnett et al. 1994; Laneuville et al. 1994), the two PGHS isoforms are remarkably similar. Both are homodimeric (~72 kDa/subunit), heme-containing, glycosylated proteins with independent but interactive peroxidase and cyclo-oxygenase (COX) active sites (Dietz et al. 1988; Smith et al. 1996). Moreover, the enzymes are novel integral membrane proteins, both of which are anchored to the luminal surfaces of the endoplasmic reticulum and both the inner and outer membranes of the nuclear envelope (Otto and Smith 1994; Spencer et al. 1998). This interaction involves the hydrophobic surfaces of amphipathic helices and only one leaflet of the lipid bilayer (Picot et al. 1994; Otto and Smith 1996). Both PGHS isozymes are important pharmacologically, because they are the major therapeutic targets of aspirin and other non-steroidal anti-inflammatory drugs (NSAIDs). Common NSAIDs inhibit both isoforms (Meade et al. 1993; Mitchell et al. 1993; Barnett et al. 1994; Laneuville et al. 1994; O'Neill et al. 1994; Patrignani et al. 1994). However, PGHS-2 (COX-2)-selective inhibitors have recently been developed (Riendeau et al. 1997; Zhang et al. 1997; Smith et al. 1998). PGHS-1 is important in thrombosis, and its inhibition by aspirin acting on platelet cells is of cardiovascular benefit (Oates et al. 1988; Patrono et al. 1990; Funk et al. 1991; Willard et al. 1992; Patrignani et al. 1994; Patrono 1994). Inhibition of PGHS-2 is anti-inflammatory, analgesic and anti-pyretic (Riendeau et al. 1997; Zhang et al. 1997; Smith et al. 1998) and may prevent colon cancer (Thun et al. 1991; Levy 1997; Tsujii et al. 1998) and Alzheimer's disease (Breitner 1996).

## 3.2 COX and Peroxidase Catalysis

PGHSs catalyze two reactions: a COX (bis-oxygenase) reaction in which arachidonate is converted to $PGG_2$ and a peroxidase reaction in which $PGG_2$ undergoes a two-electron reduction to $PGH_2$ (Fig. 1). PGHS-1 and PGHS-2 have similar COX turnover numbers (~3500 mol arachidonate/min/mol dimer; Barnett et al. 1994) and $K_m$ values for

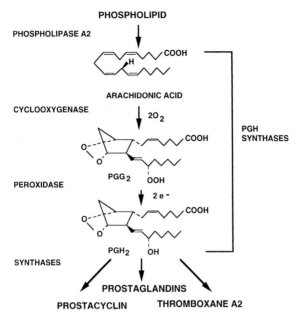

**Fig. 1.** Cyclooxygenase and peroxidase reactions catalyzed by prostaglandin endoperoxide H synthases

arachidonate (~2–5 µM; Barnett et al. 1994) and $O_2$ (5–10 µM; Lands et al. 1978; Juranek et al. 1999). However, there are subtle differences in peroxide requirements (Capdevila et al. 1995; Kulmacz and Wang 1995) and fatty-acid substrate specificities (Laneuville et al. 1995). The peroxidase activities of the two isozymes also have similar kinetic properties (Barnett et al. 1994; Laneuville et al. 1994; Gierse et al. 1995). Key amino acid residues involved in catalysis are conserved between the isozymes, and the crystal structures of the two isozymes are essentially superimposable (Picot et al. 1994; Kurumbail et al. 1996; Luong et al. 1996).

The initial and rate-limiting step in COX catalysis is abstraction of the 13-pro-S hydrogen from arachidonate to yield an arachidonyl radical (Fig. 1; Tsai et al. 1995). This is followed by sequential oxygen additions at C-11 and C-15 to yield the prostaglandin endoperoxide $PGG_2$. The peroxidase activity reduces the 15-hydroperoxide group of $PGG_2$ to an alcohol, yielding $PGH_2$. NSAIDs compete directly with arachidonate for binding to the COX site and inhibit COX activity but have no effect on peroxidase activity (Mizuno et al. 1982). The 13-pro-S hydrogen is removed from arachidonate by a tyrosyl radical (Tyr385) present in the COX active site (Dietz et al. 1988; Shimokawa et al. 1990; Picot et al. 1994; Tsai et al. 1994, 1995). Formation of the tyrosyl radical begins with a hydroperoxide-initiated oxidation of the heme group at the peroxidase active site. This two-electron oxidation of the heme group yields a peroxidase spectral intermediate (intermediate I) containing an oxyferryl form of iron (Fe IV) and a protoporphyrin radical cation (Lambeir et al. 1985; Dietz et al. 1988). The oxidized heme group oxidizes Tyr385 to yield a peroxidase intermediate (intermediate II) having a Fe IV (Dietz et al. 1988) and a tyrosyl radical (Karthein et al. 1988).

## 3.3 COX Substrates

Kinetic constants for the oxygenation of various fatty-acid substrates by human PGHS-1 and -2 are presented in Table 1. The COX activities of PGHS-1 and -2 can oxygenate a variety of C-18 and C-20 n-3 and n-6 fatty acids, including 18:2n-6, 18:3n-6, 18:3n-3, 20:2n-6, 20:3n-6, 20:4n-6 and 20:5n-3 (Lands et al. 1973; Laneuville et al. 1995). Other fatty acids, such as 18:1n-9 and 22:6n-3, are competitive inhibitors (Lands et al. 1973; Marshall and Kulmacz 1988; Laneuville et al. 1994) and are not oxygenated. Arachidonic (20:4n-6) and γ-homo-linolenic (20:3n-6) acids are utilized most efficiently by both PGHS-1 and PGHS-2, while n-3 and n-9 polyunsaturated fatty acids containing 18–22 carbons are relatively poor substrates (Lands et al. 1973; Laneuville et al. 1995). Characterization of the products formed from each of the substrates indicates that it is the n-8 allylic hydrogen which is abstracted; the exception is α-linolenate (18:3n-3), where the n-5 hydrogen is abstracted (Laneuville et al. 1995). In the case of human PGHS-1, γ-linolenic acid is a particularly poor substrate (Laneuville et al. 1995).

Figure 2 is a stereo view of arachidonic acid modeled at the ovine PGHS-1 active site. Although arachidonate can assume some $10^7$ low-energy conformations, this substrate is modeled:

**Table 1.** Fatty acid substrate specificities of human prostaglandin endoperoxide H synthase (hPGHS)-1 and -2

| Fatty acid | hPGHS-1 | | hPGHS-2 | | Relative $V_{max}/K_m$ (PGHS-2/PGHS-1) |
|---|---|---|---|---|---|
| | $K_m$ (μM) | $V_{max}$ (%) | $K_m$ (μM) | $V_{max}$ (%) | |
| Arachidonic acid (20:4n-6) | 5.4 | 100 | 5.6 | 100 | 1.0 |
| Eicosatrienoic acid (20:3n-6) | 14 | 70 | 28 | 115 | 1.2 |
| Eicosapentaenoic acid (20:5n-3) | 1.8 | 18 | 2.0 | 21 | 1.0 |
| Linoleic acid (18:2n-6) | 12 | 10 | 27 | 72 | 3.2 |
| α-Linolenic acid (18:3n-3) | 200 | 7.2 | 48 | 70 | 41 |

$K_m$ Michaelis constant; $V_{max}$ maximum rate of reaction.

1. In an overall kinked conformation appropriate for formation of a bond between C-8 and C-12, approximating its expected cyclizable structure at the time of hydrogen extraction and allowing for the entry of $O_2$ at C-11 and C-15 from the side opposite that of hydrogen abstraction (Smith and Marnett 1994; Xiao et al. 1997)
2. With the carboxylate group bound to the guanido group of Arg120 (Picot et al. 1994; Mancini et al. 1995; Bhattacharyya et al. 1996)
3. With the 13-pro-S-hydrogen neighboring the phenolic group of Tyr385 (Shimokawa et al. 1990; Picot et al. 1994)
4. With a bend in the chain at C-15, based on the crystal structure of arachidonate complexed with murine apo-PGHS-2 (Stegeman et al. 1998).

Determining the structure and configuration of arachidonate in the active site and the role of various residues in maintaining this structure will require a combination of biophysical techniques, including X-ray crystallography and $^{13}$C-nuclear magnetic resonance of arachidonate/PGHS-1 complexes and mutagenic analyses of various COX active-site residues. Here we summarize preliminary data that come

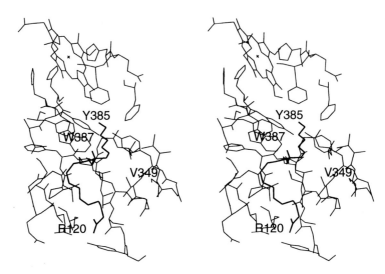

**Fig. 2.** Stereo view of arachidonic acid modeled into the cyclooxygenase active site of ovine prostaglandin endoperoxide H synthase

from mutagenic analyses of PGHS-1 and PGHS-2. Briefly, we have observed:

1. That native COX "sees" arachidonate not as a single substrate but as four different substrates (with different $K_m$ and $V_{max}$ values) that are converted into different products
2. That some mutated forms of COX see these four substrate forms differently than the native enzyme and, as a result, generate products in different proportions
3. That PGHS-1 and -2 employ different sets of active-site residues to anchor and align arachidonate and other fatty-acid substrates.

### 3.4 Native PGHS-1

A careful analysis of products formed from arachidonate by ovine PGHS-1 was performed at a variety of enzyme/substrate ratios and at different substrate concentrations (2–100 µM), using both purified and microsomal enzyme preparations. A combination of radio thin-layer chromatography, reverse-phase high-performance liquid chromatography (HPLC) and chiral HPLC indicated that four different products were formed: $PGG_2$, 11-R-hydroperoxyeicosatetraenoic acid (11-R-HPETE), 15-S-HPETE and 15-R-HPETE. At low substrate/enzyme ratios, $PGG_2$ represents greater than 95% of the product whereas, at higher substrate/enzyme ratios, the monohydroperoxy acids can be up to 40% of the total. Most importantly, it was found that the $K_m$ values for the formation of these different products are different, with approximate values of 4 μM for $PGG_2$, 8 μM for 11-R-HPETE and 12 μM for both 15-S-HPETE and 15-R-HPETE. These results indicate that there are three (or probably four) different conformations of arachidonate that can bind at the COX active site of ovine PGHS-1, each of which leads to a different product.

## 3.5 Mutant PGHS-1

We have prepared approximately 50 different mutations in the ovine PGHS-1 active site. With the exception of Arg120 mutations, which are discussed below, the mutations affect $K_m$ values by at most 10- to 20-fold. This suggests that none of these residues is essential for binding arachidonate. Instead the residues lining the hydrophobic channel seem to be most important in positioning arachidonate in the conformation that leads to $PGG_2$. For example, V349A and W387F mutants of ovine PGHS-1 lead to an excess of 11-R-HPETE (40–50% of total products), and related results were obtained for mutations at positions 530 and 518.

## 3.6 Arachidonate Binding to PGHS-1 Versus PGHS-2

In ovine PGHS-1, Arg120 appears to form an ionic linkage with arachidonate. This type of interaction was initially suggested from observation of the crystal structure (Picot et al. 1994) and is supported by the finding that an R120Q ovine PGHS-1 mutant has a $K_m$ for arachidonate approximately 1000-fold higher than that of the native enzyme (Bhattacharyya et al. 1996). Somewhat surprisingly, the homologous residue in PGHS-2 does not form an ionic linkage but instead appears to hydrogen bond to the carboxylate of arachidonate (Rieke et al. 1999). R120Q human PGHS-2 and native human PGHS-2 have nearly identical kinetic properties, whereas an R120L human PGHS-2 is considerably less active. The results of studies on Arg120 mutants of PGHS-1 and PGHS-2 imply that interactions involved in the binding of arachidonate to PGHS-1 and -2 are quite different and that residues within the hydrophobic COX channel must contribute more significantly to arachidonate binding to PGHS-2 than to PGHS-1.

## 3.7 Future Studies

Studies are currently in progress to examine the model depicted in Fig. 2. We are currently examining crystals of arachidonate complexed to a $Co^{3+}$ heme containing ovine PGHS-1. In addition, we are testing different n-6 and n-3 fatty acids as substrates for both native and mutant ovine PGHS-1.

**Acknowledgments.** This work was supported in part by P01-GM57323 and R01-DK22042 from the National Institutes of Health.

## References

Barnett J, Chow J, Ives D, Chiou M, Mackenzie R, Osen E, Nguyen B, Tsing S, Bach C, Freire J, et al. (1994) Purification, characterization and selective inhibition of human prostaglandin G/H synthase 1 and 2 expressed in the baculovirus system. Biochim Biophys Acta 1209:130–139

Bhattacharyya DK, Lecomte M, Rieke CJ, Garavito RM, Smith WL (1996) Involvement of arginine 120, glutamate 524, and tyrosine 355 in the binding of arachidonate and 2-phenylpropionic acid inhibitors to the cyclooxygenase active site of ovine prostaglandin endoperoxide H synthase-1. J Biol Chem 271:2179–2184

Breitner JC (1996) Inflammatory processes and anti-inflammatory drugs in Alzheimer's disease: a current appraisal. Neurobiol Aging 17:789–794

Capdevila JH, Morrow JD, Belosludtsev YY, Beauchamp DR, DuBois RN, Falck JR (1995) The catalytic outcomes of the constitutive and the mitogen inducible isoforms of prostaglandin H2 synthase are markedly affected by glutathione and glutathione peroxidase. Biochemistry 34:3325–3337

Dietz R, Nastainczyk W, Ruf HH (1988) Higher oxidation states of prostaglandin H synthase. Rapid electronic spectroscopy detected two spectral intermediates during the peroxidase reaction with prostaglandin G2. Eur J Biochem 171:321–328

Funk CD, Funk LB, Kennedy ME, Pong AS, Fitzgerald GA (1991) Human platelet/erythroleukemia cell prostaglandin G/H synthase: cDNA cloning, expression, and gene chromosomal assignment. FASEB J 5:2304–2312

Gierse JK, Hauser SD, Creely DP, Koboldt C, Rangwala SH, Isakson PC, Seibert K (1995) Expression and selective inhibition of the constitutive and inducible forms of human cyclo-oxygenase. Biochem J 305:479–484

Juranek I, Suzuki H, Yamamoto S (1999) Affinities of various mammalian arachidonate lipoxygenases and cyclooxygenases for molecular oxygen as substrate. Biochim Biophys Acta 1436:509–518

Karthein R, Dietz R, Nastainczyk W, Ruf HH (1988) Higher oxidation states of prostaglandin H synthase. EPR study of a transient tyrosyl radical in the enzyme during the peroxidase reaction. Eur J Biochem 171:313–320

Kulmacz RJ (1998) Cellular regulation of prostaglandin H synthase catalysis. FEBS Lett 430:154–157

Kulmacz RJ, Wang L-H (1995) Comparison of hydroperoxide initiator requirements for the cyclooxygenase activities of prostaglandin H synthase-1 and -2. J Biol Chem 270:24019–24023

Kurumbail RG, Stevens AM, Gierse JK, McDonald JJ, Stegeman RA, Pak JY, Gildenaus D, Miyashiro JM, Penning TD, Seibert K, Isakson PC, Stallings WC (1996) Structural basis for selective inhibition of cyclooxygenase-2 by anti-inflammatory agents. Nature 384:644–648

Lambeir AM, Markey CM, Dunford HB, Marnett LJ (1985) Spectral properties of the higher oxidation states of prostaglandin H synthase. J Biol Chem 260:14894–14896

Lands WEM, LeTellier PR, Rome LH, Vanderhoek JY (1973) Inhibition of prostaglandin biosynthesis. Adv Biosci 9:15–28

Lands WEM, Sauter J, Stone GW (1978) Oxygen requirement for prostaglandin biosynthesis. Prostaglandins Med 1:117–120

Laneuville O, Breuer DK, DeWitt DL, Hla T, Funk CD, Smith WL (1994) Differential inhibition of human prostaglandin endoperoxide H synthases-1 and -2 by nonsteroidal anti-inflammatory drugs. J Pharmacol Exp Ther 271:927–934

Laneuville O, Breuer DK, Xu N, Huang ZH, Gage DA, Watson JT, Lagarde M, DeWitt DL, Smith WL (1995) Fatty acid substrate specificities of human prostaglandin endoperoxide H synthases-1 and -2. Formation of 12 hydroxy-(9Z, 13E/Z, 15Z)-octadecatrienoic acids from α-linolenic acid. J Biol Chem 270:19330–19336

Levy GN (1997) Prostaglandin H synthases, nonsteroidal anti-inflammatory drugs, and colon cancer. FASEB J 11:234–247

Luong C, Miller A, Barnett J, Chow J, Ramesha C, Browner MF (1996) Flexibility of the NSAID binding site in the structure of human cyclooxygenase-2. Nat Struct Biol 3:927–933

Mancini JA, Riendeau D, Falgueyret JP, Vickers PJ, O'Neill GP (1995) Arginine 120 of prostaglandin G/H synthase-1 is required for the inhibition by nonsteroidal anti-inflammatory drugs containing a carboxylic acid moiety. J Biol Chem 270:29372–29377

Marshall PJ, Kulmacz RJ (1988) Prostaglandin H synthase: distinct binding sites for cyclooxygenase and peroxidase substrates. Arch Biochem Biophys 266:162–170

Meade EA, Smith WL, DeWitt DL (1993) Differential inhibition of prostaglandin endoperoxide synthase (cyclooxygenase) isozymes by aspirin and other non-steroidal anti-inflammatory drugs. J Biol Chem 268:6610–6614

Mitchell JA, Akarasereenont P, Thiemermann C, Flower RJ, Vane JR (1993) Selectivity of nonsteroidal anti-inflammatory drugs as inhibitors of constitutive and inducible cylooxygenase. Proc Natl Acad Sci U S A 90:11693–11697

Mizuno K, Yamamoto S, Lands WEM (1982) Effects of non-steroidal anti-inflammatory drugs on fatty acid cyclooxygenase and prostaglandin hydroperoxidase activities. Prostaglandins 23:743–757

Oates JA, Fitzgerald GA, Branch RA, Jackson EK, Knapp HR, Roberts LJ (1988) Clinical implications of prostaglandin and thromboxane A2 formation. N Engl J Med 319:689–698

O'Neill GP, Mancini JA, Kargman S, Yergey J, Kwan MY, Falgueyret JP, Abramovitz M, Kennedy BP, Ouellet M, Cromlish W, et al. (1994) Overexpression of human prostaglandin G/H synthase-1 and -2 by recombinant vaccinia virus: inhibition by nonsteroidal anti-inflammatory drugs and biosynthesis of 15-hydroxyeicosatetraenoic acid. Mol Pharmacol 45:245–254

Otto JC, Smith WL (1994) The orientation of prostaglandin endoperoxide synthases-1 and -2 in the endoplasmic reticulum. J Biol Chem 269:19868–19875

Otto JC, Smith WL (1996) Photolabeling of prostaglandin endoperoxide H synthase-1 with 3-trifluoro-3-(m-[$^{125}$I]-iodophenyl)diazirine as a probe of membrane association and the cyclooxygenase active site. J Biol Chem 271:9906–9910

Patrignani P, Panara MR, Greco A, Fusco O, Natoli C, Iacobelli S, Cipollone F, Ganci A, Creminon C, Maclouf J, et al. (1994) Biochemical and pharmacological characterization of the cyclooxygenase activity of human blood prostaglandin endoperoxide synthases. J Pharmacol Exp Ther 271:1705–1712

Patrono C (1994) Aspirin as an antiplatelet drug. N Engl J Med 330:1287–1294

Patrono C, Ciabattoni G, Davi G (1990) Thromboxane biosynthesis in cardiovascular diseases. Stroke 21:IV130–IV133

Picot D, Loll PJ, Garavito M (1994) The X-ray crystal structure of the membrane protein prostaglandin H2 synthase-1. Nature 367:243–249

Rieke CJ, Mulichak AM, Garavito RM, Smith WL (1999) The role of Arg120 of prostaglandin endoperoxide H synthase-2 in the interaction with fatty acid substrates and inhibitors. J Biol Chem 274:17109–17114

Riendeau D, Percival MD, Boyce S, Brideau C, Charleson S, Cromlish W, Ethier D, Evans J, Falgueyret JP, Ford-Hutchinson AW, Gordon R, Greig G, Gresser M, Guay J, Kargman S, Leger S, Mancini JA, O'Neill G, Ouellet M, Rodger IW, Therien M, Wang Z, Webb JK, Wong E, Chan CC, et al. (1997) Biochemical and pharmacological profile of a tetrasubstituted furanone as a highly selective COX-2 inhibitor. Br J Pharmacol 121:105–117

Shimokawa T, Kulmacz RJ, DeWitt DL, Smith WL (1990) Tyrosine 385 of prostaglandin endoperoxide synthase is required for cyclooxygenase catalysis. J Biol Chem 265:20073–20076

Smith CJ, Zhang Y, Koboldt CM, Muhammad J, Zweifel BS, Shaffer A, Talley JJ, Masferrer JL, Seibert K, Isakson PC (1998) Pharmacological analysis of cyclooxygenase-1 in inflammation. Proc Natl Acad Sci U S A 95:13313–13318

Smith WL, DeWitt DL (1996) Prostaglandin endoperoxide H synthases-1 and -2. In: Dixon FJ (ed) Advances in immunology, vol 62. Academic, Orlando, pp 167–215

Smith WL, Marnett LJ (1994) Prostaglandin endoperoxide synthases. In: Sigel H, Sigel A (eds) Metal ions in biological systems, vol 30. Dekker, New York, pp 163–199

Smith WL, Garavito RM, DeWitt DL (1996) Prostaglandin endoperoxide H synthases (cyclooxygenases)-1 and -2. J Biol Chem 271:33157–33160

So O-Y, Scarafia LE, Mak AY, Callan OH, Swinney DC (1998) The dynamics of prostaglandin H synthases. Studies with prostaglandin H synthase 2 Y355F unmask mechanisms of time-dependent inhibition and allosteric activation. J Biol Chem 273:5801–5807

Spencer AG, Woods JW, Arakawa T, Singer II, Smith WL (1998) Subcellular localization of prostaglandin endoperoxide H synthases-1 and -2 by immunoelectron microscopy. J Biol Chem 273:9886–9893

Stegeman R, Pawlitz J, Stevens A, Gierse J, Stallings W, Kurumbail R (2000) Mechanism of cyclooxygenase reactions: structure of arachidonic acid bound to cyclooxygenase-2. Acta Crystallogr C (in press)

Thun MJ, Namboodiri MM, Heath CW (1991) Aspirin use and reduced risk of fatal colon cancer. N Engl J Med 325:1593–1596

Tsai A, Hsi LC, Kulmacz RJ, Palmer G, Smith WL (1994) Characterization of the tyrosyl radicals in ovine prostaglandin H synthase-1 by isotope replacement and site-directed mutagenesis. J Biol Chem 269:5085–5091

Tsai A, Kulmacz RJ, Palmer G (1995) Spectroscopic evidence for reaction of prostaglandin H synthase-1 tyrosyl radical with arachidonic acid. J Biol Chem 270:10503–10508

Tsujii M, Kawano S, Tsuji S, Sawaoka H, Hori M, DuBois RN (1998) Cyclooxygenase regulates angiogenesis induced by colon cancer cells. Cell 93:705–716

Willard J, Lange RA, Hillis LD (1992) The use of aspirin in ischemic heart disease. N Engl J Med 327:175–181

Xiao G, Tsai AL, Palmer G, Boyar WC, Marshall PJ, Kulmacz RJ (1997) Analysis of hydroperoxide-induced tyrosyl radicals and lipoxygenase activity in aspirin-treated human prostaglandin H synthase-2. Biochemistry 36:1836–1845

Zhang Y, Shaffer A, Portanova J, Seibert K, Isakson PC (1997) Inhibition of cyclooxygenase-2 rapidly reverses inflammatory hyperalgesia and prostaglandin E2 production. J Pharmacol Exp Ther 283:1069–1075

# 4 Structure, Function and Inhibition of Cyclo-oxygenases

L. J. Marnett

4.1 Introduction ............................................. 65
4.2 COX Catalysis .......................................... 66
4.3 COX Inhibition ......................................... 70
4.4 Perspective ............................................. 75
References ................................................... 77

## 4.1 Introduction

The key step in the generation of prostaglandins (PGs) is the bis-dioxygenation of arachidonic acid to the bicyclic hydroperoxides $PGG_2$ (endoperoxides) as the initial products of polyunsaturated fatty-acid oxygenation (Fig. 1). The enzyme that catalyzes this reaction is cyclo-oxygenase (COX)[1], and it also catalyzes the reduction of the hydroperoxide to an alcohol, thereby forming $PGH_2$ (Hamberg and Samuelsson 1973; Nugteren and Hazelhof 1973).

In addition to catalyzing a fascinating metabolic transformation, COX is also an enormously important pharmacological target. In 1971, Vane reported that non-steroidal anti-inflammatory drugs (NSAIDs) inhibit PG formation and demonstrated that their relative inhibitory potency in vitro correlates to their anti-inflammatory activity in vivo

---

[1] The term "cyclo-oxygenase" is used to describe the enzyme or to refer to the active site of the protein.

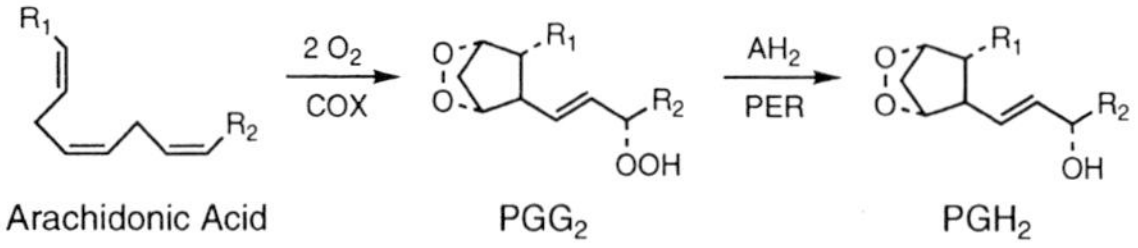

**Fig. 1.** Arachidonic-acid oxygenation by cyclo-oxygenase and reduction of prostaglandin $G_2$ by peroxidase

$R_1 = CH_2CH=CH(CH_2)_3CO_2H$; $R_2 = C_5H_{11}$; $AH_2$ = reducing substrate

(Vane 1971). This explained both the beneficial activity of NSAIDs and their side effects (such as gastrointestinal toxicity and bleeding) because PGs and related molecules (i.e., thromboxanes) are involved in a very broad range of physiological and pathophysiological responses. The importance of these molecules as autocrine and paracrine mediators has been confirmed recently by the phenotypes of mice bearing targeted deletions in COX genes or PG-receptor genes.

The discovery of a second gene (COX-2) coding for COX and the demonstration that its protein product is distributed differently from the originally discovered enzyme (COX-1) raised the possibility that it may be possible to separate some of the beneficial effects of NSAIDs from their side effects by developing isoform-selective inhibitors (Fu et al. 1990; Kujubu et al. 1991; O'Banion et al. 1991; Xie et al. 1991). This hypothesis has been dramatically validated by the demonstration that selective COX-2 inhibitors are anti-inflammatory and analgesic but lack the gastric toxicity associated with all currently available NSAIDs (Masferrer et al. 1994; Simon et al. 1998). These discoveries have refocused attention on the structure and function of COX enzymes and their inhibition.

## 4.2 COX Catalysis

Substantial evidence supports the hypothesis that COX oxygenates arachidonic acid by a free-radical mechanism (Fig. 2). Thus, to generate PGs, COX appears to have co-opted the process that gives rise to isoprostanes. The major differences between COX-catalyzed and spontaneous oxidation of arachidonic acid is the increased rate of the enzy-

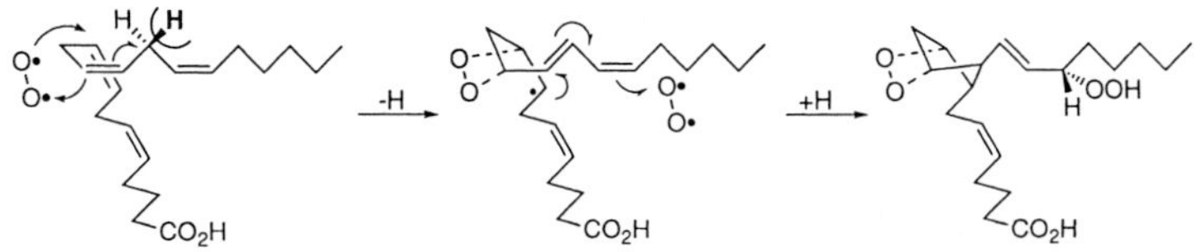

**Fig. 2.** Chemical steps in the conversion of arachidonic acid to prostaglandin $G_2$ ($PGG_2$). The enzyme removes the 13-pro-(S)-hydrogen, which generates a pentadienyl radical with maximun electron density at C-11 and C-15. Trapping of the carbon radical at C-11 with $O_2$ produces a peroxyl radical, which adds to C-9, generating a cyclic peroxide and a carbon-centered radical at C-8. The C-8 radical adds to the double bond at C-12, generating the bicyclic peroxide and an allylic radical with maximum electron density at C-13 and C-15. Trapping of the carbon radical at C-15 radical with $O_2$ generates a peroxyl radical, which is reduced to $PGG_2$

matic reaction and its high degree of stereochemical control (1 of 64 possible isomers predominates). Thus, the overall roles of COX are rather simple: it stereospecifically removes the 13-pro-*S*-hydrogen and controls the stereochemistry of oxygenation.

### 4.2.1 Oxidizing Agent

A protein tyrosyl radical appears to be the oxidizing agent that initiates arachidonic-acid oxygenation (Karthein et al. 1988). A tyrosyl radical is formed during COX turnover and, although there has been debate over the identity of the spectroscopically detected radicals, they appear to be capable of oxidizing arachidonic acid (DeGray et al. 1992; Tsai et al. 1992, 1998). The crystal structures of both COX-1 and COX-2 reveal that Tyr385 is positioned perfectly to react with the fatty-acid substrate (Picot et al. 1994; Kurumbail et al. 1996; Luong et al. 1996). Indeed, the Tyr385Phe mutant is not catalytically active and does not oxidize arachidonic acid when it is treated with peroxide (Shimokawa et al. 1990). Incubation of wild-type enzyme with arachidonate in the presence of nitric oxide (NO) quenches the electron paramagnetic resonance signal of the tyrosyl radical and leads to the formation of nitrotyrosine at position 385 in the protein (Gunther et al. 1997; Goodwin et al. 1998).

Protein radicals require an oxidant for their formation; in most cases, this oxidant is a metal-containing prosthetic group (Stubbe and Van der Donk 1998). COX is a homodimer of 70-kDa subunits that each contain one molecule of heme (Picot et al. 1994). The iron is ferric in the resting enzyme and is likely thermodynamically incapable of oxidizing Tyr385 ($E_{1/2}$=0.9 V for Tyr•→Tyr and $E_{1/2}$=–0.2 to +0.2 V for $Fe^{3+}$→$Fe^{2+}$ for most hemes; Kulmacz et al. 1990; Stubbe and Van der Donk 1998). Reaction of the heme of COX with peroxides generates a ferryl–oxo complex analogous to Compound I of classic heme peroxidases (Lambeir et al. 1985). The redox potential of such higher oxidation states is typically on the order of +1 V, so Compound I of COX is capable of oxidizing Tyr385. Ruf and co-workers demonstrated that oxidation of COX with organic hydroperoxides or fatty-acid hydroperoxides generates a spectroscopically detectable tyrosyl radical, and they postulated that the tyrosyl radical oxidizes arachidonic acid (Dietz et al. 1988). Support for this hypothesis is provided by the existence of significant lag phases for the induction of COX activity of Mn–protoporphyrin-IX reconstituted enzyme or site-directed mutants that exhibit diminished rates of reaction with hydroperoxide (Smith et al. 1992).

The identity of the hydroperoxide activator is uncertain. Our laboratory has reported that peroxynitrite, the coupling product of NO and superoxide anion, is an excellent oxidant for the heme of COX and activates the enzyme even in the presence of concentrations of glutathione peroxidase and glutathione that inhibit activation by fatty-acid hydroperoxides (Landino et al. 1996). Activation is inhibited by superoxide dismutase, which scavenges peroxynitrite or prevents its formation from NO and $O_2^-$. Lipophilic superoxide dismutase-mimetic agents inhibit PG biosynthesis by cultured mouse macrophages, which is consistent with a role for peroxynitrite in COX activation in intact cells. These findings provide a biochemical link between NO biosynthesis and PG biosynthesis and may explain the finding that NO-synthase inhibitors reduce PG biosynthesis in inflammatory lesions in vivo (Eq. 1) (Salvemini et al. 1995). Peroxynitrite activation of COX may be especially important in activated macrophages, because inducible NO synthase and COX-2 are immediate early genes that are dramatically expressed in response to exposure to inflammatory stimuli, such as lipopolysaccharide. The identity of the COX activator in non-inflammatory cells remains to be determined.

$$\cdot N{=}O \; + \; \cdot O{-}O^{\ominus} \;\xrightarrow{6 \times 10^9 \, M^{-1}s^{-1}}\; O{=}N{-}O{-}O^{\ominus} \quad (1)$$

Nitric Oxide  Superoxide           Peroxynitrite

### 4.2.2 Stereochemistry

How does COX ensure that a single stereoisomer of $PGG_2$ is produced from arachidonic acid? One can predict on purely chemical grounds that the enzyme must bind arachidonate in a conformation similar to that illustrated in Fig. 2. Removal of the 13-pro-$S$-hydrogen followed by reaction with $O_2$, serial cyclization and reaction with the second $O_2$ could occur (with minimal motion of the reaction intermediates) to produce $PGG_2$ with all the correct stereocenters. We have docked arachidonate to the cyclo-oxgenase active site of sheep COX-1 to test whether such a conformation can be accommodated (Rowlinson et al. 1999). The carboxylate was positioned adjacent to Arg120, which plays a crucial role in binding arachidonate and arylalkanoic acid-type inhibitors (Bhattacharyya et al. 1996), and the 13-pro-$S$-hydrogen was placed near the phenolic hydroxyl of Tyr385. The ω-end of the fatty acid was inserted into a channel at the top of the COX active site that is eventually located at the surface of the protein. The validity of the model is suggested by the observation that site-directed mutation of Gly533 into Val or Leu abolishes oxygenation of the 20-carbon fatty acid (arachidonic acid) but not the oxygenation of the 18-carbon fatty acid (α-linolenic acid; Rowlinson et al. 1999). This end of arachidonate straddles the α-helix containing Ser530, the site acetylated by aspirin. Acetylation by aspirin blocks arachidonate binding to COX-1 (DeWitt et al. 1990).

This model was developed using coordinates for sheep COX-1, but a similar model can be developed using the coordinates for mouse COX-2. The active-site structures are quite similar for the two forms of the enzyme, with a few exceptions (detailed below). Despite this similarity, clear differences in substrate specificities exist between the two enzymes. COX-2 appears to be much more accommodating than COX-1 in that it oxidizes 18-carbon polyunsaturated fatty acids with much higher efficiency than COX-1 does. Also, COX-2 oxidizes the hydroxyethylamide derivative of arachidonic acid (anandamide), whereas COX-1 does not (Laneuville et al. 1995; Yu et al. 1997). Consistent with the latter observation, the Arg120Gln mutant of COX-1 demonstrates a 100-fold increase in $K_m$ for arachidonate, whereas the corresponding mutant of COX-2 displays a $K_m$ for arachidonate that is quite similar to that for the wild-type enzyme (Bhattacharyya et al. 1996; Rieke et al. 1999). The molecular basis for these differences is not understood. Another difference between the two enzymes is the ability of aspirin-acetylated COX-2 to oxygenate arachidonic acid to 15-R-hydroxyeicosatetraenoic acid (Lecomte et al. 1994; Xiao et al. 1997). Acetylated COX-1 is unable to carry out this transformation.

## 4.3 COX Inhibition

### 4.3.1 Classes of Selective COX-2 Inhibitors

Enormous effort was expended in the development of NSAIDs between the 1960s and 1980s, so many pharmacophores were available for testing when COX-2 was discovered (for a historical review of the medicinal chemistry of COX-2 inhibitors, see Talley 1999). The first breakthrough came with reports that the diaryl heterocycle DuP-697 and the acidic sulfonanilide NS-398 were anti-inflammatory but not ulcerogenic (Gans et al. 1990; Futaki et al. 1994). Once it was demonstrated that they were COX-2-selective inhibitors, the race was on to elaborate them into structurally related clinical candidates. The COX-2 inhibitors currently on the market, celebrex and vioxx, are descendants in the diaryl-heterocycle lineage (Fig. 3). Structure–activity studies indicate that a *cis*-stilbene moiety containing a 4-methylsulfonyl or sulfonamide substituent in one of the pendant phenyl rings is required for COX-2 specificity (Talley et al. 1999). The oxidation state of the sulfur in the methylsulfone moiety is critical for selective COX-2 inhibition, because the sulfoxides or sulfides are inactive or non-selective. The ring system that is fused to the stilbene framework has been extensively manipulated to include heterocyclic, fused-heterocyclic and carbocyclic templates (Prasit and Riendeau 1997; Talley 1999).

Multiple strategies have been employed to build COX-2 selectivity into the powerful NSAID indomethacin (Fig. 3; Black et al. 1996; Takeuchi et al. 1998). These include lengthening the alkyl–carboxylic-

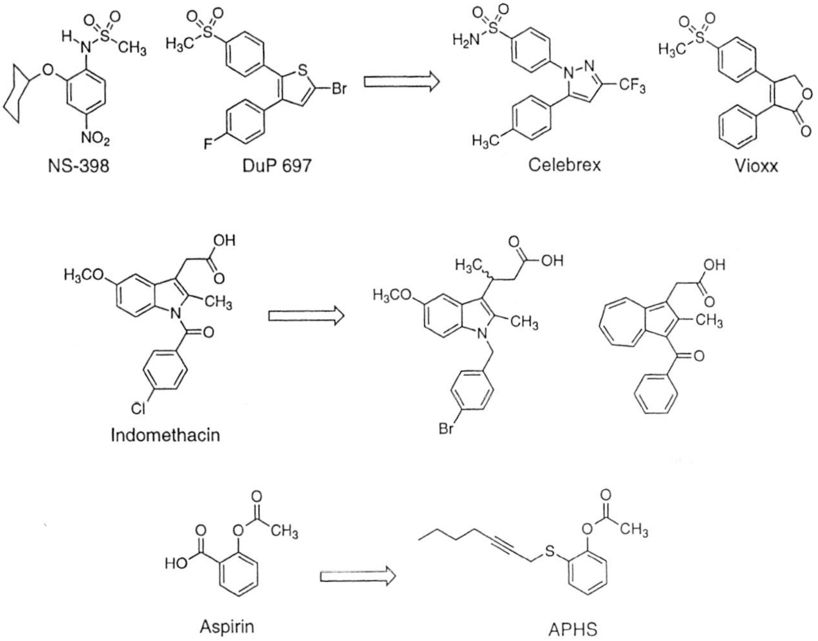

**Fig. 3.** Different classes of cyclo-oxygenase-2 inhibitors

acid side chain, modifying the aroyl group attached to the indole nitrogen and transforming the indole ring into an azulene functionality. Although compounds with reasonable in vitro and in vivo properties have been reported, none of them appears to have been advanced as a drug candidate. Efforts to convert other NSAIDs into COX-2-selective inhibitors have been reported but, in the interest of space, are not cited.

Our laboratory recently reported the development of irreversible inactivators of COX-2 that act like aspirin (Kalgutkar et al. 1998). Aspirin is the only COX inhibitor that covalently modifies COX enzymes, but it is more potent against COX-1 than against COX-2 (Roth et al. 1975). This feature accounts for its anti-platelet/cardiovascular effects and its ulcerogenic side effects. We found that replacement of aspirin's carboxylic acid with alkylsulfide or alkynylsulfide functionalities generates COX-2-selective inactivators that acetylate the same serine residue as aspirin [*o*-(acetoxyphenyl)hept-2-ynyl sulfide (APHS); Fig. 2; Kalgutkar et al. 1998; Marnett and Kalgutkar 1998]. APHS is 15- to 20-fold more active than aspirin in the rat-paw edema model of inflammation but does not induce gastric ulcers during acute testing. The possibility that irreversible inactivation of COX-2 will provide some added benefit relative to reversible inhibitors is currently under investigation.

### 4.3.2 Mechanism of Selective COX-2 Inhibition

It has been known for many years that inhibition of COX by most NSAIDs conforms to a minimal two-step binding mechanism (Eq. 2) (Rome and Lands 1975). The first step involves the formation of a readily reversible complex (E·I) that represents a competitively inhibited enzyme. Some NSAIDs, such as ibuprofen and mefenamic acid, are purely competitive inhibitors of COX-1 and COX-2. The second step involves the conversion of the (E·I) complex to the (E·I*) complex, in which the inhibitor is bound more tightly to the enzyme. This step occurs in seconds or minutes and may reflect the induction of a protein conformational change. Time-dependent inhibitors include indomethacin, naproxen and meclofenamic acid *inter alia*. Interestingly, all of the selective COX-2 inhibitors inhibit COX-1 and COX-2 competitively but are time-dependent only against COX-2 (i.e., COX-2 selectivity is dependent on the time-dependent step; Copeland et al. 1994).

$$E + I \underset{k_{-1}}{\overset{k_1}{\rightleftharpoons}} (E \cdot I) \xrightarrow{k_2} (E \cdot I)^* \qquad (2)$$

This complex kinetic mechanism of inhibition, along with normal lab-to-lab idiosyncrasies in the design of enzyme assays, has made it very difficult to compare COX-2 selectivity ratios for compounds reported in the literature (Gierse et al. 1999). Since all COX-2-selective inhibitors are slow, tight-binding inhibitors that competitively inhibit both enzymes, the selectivity observed is a function of the substrate concentration. Assays that employ saturating concentrations of arachidonic acid cannot detect competitive inhibition of COX-1, so the COX-2 selectivity is high, because inhibition derives entirely from the time-dependent process. Assays that use lower concentrations of arachidonate allow some competitive inhibition, so the selectivity may be lower. Another factor complicating selectivity comparisons is the concentration of enzyme used for the in vitro assays. Many of the COX-2 inhibitors are so powerful that they are essentially one-to-one titrants of active enzyme. Consequently, the minimum obtainable $IC_{50}$ for inhibition depends on the enzyme concentration. A corollary of this is that the selectivity ratio will reflect (to some extent) the relative amounts of COX-1 and COX-2 in the assays. Our experience suggests that the best results are obtained when comparable amounts of each enzyme (based on activity) are used in the assays.

There is not general agreement on an optimal value for COX-2 selectivity. In fact, no clear correlation exists between selectivities reported from in vitro assays and those estimated from in vivo or ex vivo assays. The disparity between the selectivities measured in enzyme assays and those measured in the human whole-blood assay provides an illustrative example of this problem. The whole-blood assay determines selectivity against platelet COX-1 and monocyte COX-2 and is believed by some to be the gold standard for advancing compounds as clinical candidates. Warner et al. recently compared over 40 NSAIDs and COX-2 inhibitors in the standard version of this assay and in a modified version that utilizes supplementation of blood with activated human airway epithelial cells to provide a standardized source of COX-2 (Warner et al. 1999). In this extensive study, celebrex demonstrated 1.4-fold selectivity for COX-2 in the standard assay and fourfold selectivity in the modified assay. Both values contrast sharply with the 300-fold selectivity reported in enzyme assays (Penning et al. 1997). Based on the data from the whole-blood assay, one would expect to see significant inhibition of platelet COX-1 in clinical trials, but this does not appear to be the case (McAdam et al. 1999).

The biochemical basis for the disparities in selectivity measured in different assays remains unclear but may be related to dynamic aspects of COX-2-inhibitor action. We recently demonstrated that the rates of dissociation of selective inhibitors from COX-2 are extremely slow – in fact, several orders of magnitude slower than the rates of dissociation from COX-1 (Lanzo 1998). We hypothesize that, as metabolism reduces the levels of these selective inhibitors, they rapidly dissociate from

COX-1 but remain tightly bound to COX-2. This may give a longer duration of action against COX-2 than against COX-1, which would lead to higher selectivities than estimated using a static method of evaluation. This hypothesis should be testable in an appropriate clinical experiment.

### 4.3.3 Structural Basis for Selective COX-2 Inhibition

Several recent reviews focus on structural aspects of COX catalysis and inhibition (Smith et al. 1996; DeWitt 1999; Marnett et al. 1999). Crystal structures of ovine COX-1, murine COX-2 and human COX-2 have been solved at 3- to 3.5-Å resolution (Picot et al. 1994; Kurumbail et al. 1996; Luong et al. 1996). The overall folding patterns are very similar, which is not surprising, considering that COX isoforms are approximately 65% identical in sequence. All COX inhibitors, regardless of their selectivity, bind at the arachidonic-acid binding site. This is located in the upper half of a long channel leading from the membrane interface to the interior of the protein. A constriction formed by residues Arg120, Tyr355 and Glu524 marks the bottom of the arachidonate-binding site and restricts access to it. Opening and closing of this constriction may contribute to the time-dependence of inhibition by some NSAIDs.

Carboxylic acid-containing NSAIDs and arachidonate ion pair with the guanidinium group of Arg120, which is the only positively charged residue in the arachidonate-binding channel (Loll et al. 1995, 1996; Kurumbail et al. 1996). Site-directed mutagenesis of the Arg residue in COX-1 to Gln or glutamate renders the protein resistant to inhibition by NSAIDs and increases the $K_m$ for arachidonate binding (Mancini et al. 1995; Bhattacharyya et al. 1996). Tyr355 sterically hinders the mouth of the COX active site, which accounts for the preferential inhibition exhibited by S-stereoisomers of α-methyl-substituted arylalkanoic inhibitors (the profens; So et al. 1998).

The COX-2 selectivity of diaryl heterocycles appears to be due to the insertion of the sulfonamide or sulfone moiety into a side pocket of the main arachidonate-binding channel (Kurumbail et al. 1996). This side pocket is bordered by Val523, and the corresponding region of COX-1 is inaccessible because of the presence of an Ile residue instead of Val at position 523, which sterically hinders inhibitor approach. Other changes between COX-2 and COX-1 that contribute to rigidification of this side pocket include the substitutions Arg513His and Val434Ile. The site-directed COX-2 mutant Val523Ile is resistant to inhibition by diaryl heterocycles and acidic sulfonamides but not to alkanoic acid-type NSAIDs (Gierse et al. 1996; Guo et al. 1996; Wong et al. 1997). The reverse mutations in COX-1 render the mutant enzymes sensitive to time-dependent inhibition by diaryl heterocycles (Wong et al. 1997). Movement of Val523 may contribute to the time dependence of inhibition by diaryl heterocycles.

The structural basis for selectivity by the extended indomethacin derivatives is dependent on substitutions at the top rather than the side of the COX active site. A co-crystal structure of human COX-2 complexed with a 4-bromobenzyl indomethacin analog reveals that the 4-bromobenzyl group is in van der Waals contact with Leu503 at the apex of the COX-2 active site. Position 503 in COX-1 is substituted with Phe, which is not as easily displaced by the bromobenzyl group as Leu and may account for the selectivity of inhibition.

The structural analysis of COX-inhibitor complexes has been extremely useful in probing mechanisms of selectivity, but it did not have a major impact on the development of the first generation of COX-2 inhibitors. In fairness to the crystallographers, the speed with which diaryl heterocycles were developed rendered this practically impossible. However, structural information is now being put to use in inhibitor design, and the first examples of structure-based COX-2 inhibitors are beginning to appear (Bayly et al. 1999).

## 4.4 Perspective

Although the development of NSAIDs goes back thousands of years and, in fact, spawned the modern pharmaceutical industry, we are at the beginning of a new era in their use. After nearly 30 years of study, the true molecular target of their anti-inflammatory activity has been identified, and selective inhibitors of this target are available to the general population. Based on clinical-trial data, COX-2 inhibitors would appear to be the drugs of choice for patients with gastric sensitivity to NSAIDs (Simon et al. 1998; Ehrich et al. 1999). In fact, they may provide a new therapeutic option to patients who long ago eschewed NSAIDs.

Celebrex and vioxx do not appear to be more efficacious than current NSAIDs as anti-inflammatory agents, and they will not be safer than NSAIDs in all situations. For example, there are no data yet to indicate that NSAID-sensitive asthmatics will tolerate COX-2 inhibitors any better. Furthermore, other side effects will almost certainly emerge as individuals take higher doses of COX-2 inhibitors because of their superior gastric tolerability. COX-2 appears to play a role in wound healing, so some suggest that COX-2 inhibitors may slow this process (Wallace 1999). In addition, COX-2 inhibitors do not inhibit platelet COX-1, so they may unfavorably alter the thromboxane/prostacyclin balance by inhibiting COX-2-dependent prostacyclin formation in vascular endothelial cells (McAdam et al. 1999). Individuals with severe thrombotic disorders may be more sensitive to COX-2 inhibitors. Such adverse effects were not seen in phase-III testing, but the number of individuals in the general population taking COX-2 inhibitors will soon be three orders of magnitude higher than the numbers who participated in trials.

Celebrex and vioxx may not be more potent than current NSAIDs as anti-inflammatory agents, but they clearly have superior gastrointestinal safety profiles. Considering that the number of deaths from NSAID-induced bleeding ulcers in the United States alone is comparable to the number of deaths from acquired immune deficiency syndrome and violent crime, significant reductions in that number would be considered a major public-health accomplishment (Singh and Rosen-Ramey 1998). Much more powerful COX-2 inhibitors are undergoing human testing and may eventually become the anti-inflammatory drugs of first choice.

A key feature of the debate on the likely success of COX-2 inhibitors as anti-inflammatory and analgesic agents is the prior existence of drugs that are relatively inexpensive and safe for most people who take them (Hawkey 1999). This is not the case for some of the other clinical settings indicated for the use of COX-2 inhibitors. For example, cancer chemoprevention is in its infancy, and there are essentially no drugs that are efficacious and safe enough to take for prolonged periods of time. However, advances in molecular biology are making it possible to identify individuals at high risk for certain cancers. What can be done for these people? There is compelling epidemiological and experimental evidence to support a role for COX-2 in the progression of colon cancer and (possibly) bladder and esophageal cancer (Thun et al. 1991; DuBois et al. 1998). As a result, several trials of COX-2 inhibitors for the prevention of colon cancer are underway. Animal testing for cancer-chemopreventive activity indicates that, unlike its activity in inflammation, celebrex is far superior to any other NSAID that has been tested in the azoxymethane-induced rodent model of colon carcinogenesis (Kawamori et al. 1998). Furthermore, recent data indicate that COX-2 inhibitors are powerful anti-angiogenic agents in vivo, whereas COX-1 inhibitors are not (Tsujii et al. 1998). Thus, COX-2 inhibitors may have an adjuvant role in the treatment of tumors and a primary role in cancer prevention.

It is fitting that COX-2 inhibitors are appearing on the market at the turn of the century, approximately 100 years after the introduction of their predecessor, aspirin. As the new millennium begins, one looks forward to reaping the benefits of the molecular-biology revolution through improvements in human health. Although COX-2 inhibitors are not curative, they may improve the quality of life of an aging population afflicted with arthritis, and they may prevent the genesis of chronic diseases that are scourges of human existence.

**Acknowledgement.** Work in the Marnett laboratory is supported by a research grant from the National Institutes of Health (CA47479).

# References

Bayly CI, Black WC, Léger S, Ouimet N, Ouellet M, Percival MD (1999) Structure-based design of COX-2 selectivity into flurbiprofen. Bioorg Med Chem Lett 9:307–312

Bhattacharyya DK, Lecomte M, Rieke CJ, Garavito RM, Smith WL (1996) Involvement of arginine 120, glutamate 524, and tyrosine 355 in the binding of arachidonate and 2-phenylpropionic acid inhibitors to the cyclooxygenase active site of ovine prostaglandin endoperoxide H synthase-1. J Biol Chem 271:2179–2184

Black WC, Bayly C, Belley M, Chan C-C, Charleson S, Denis D, Gauthier JY, Gordon R, Guay D, Kargman S, Lau CK, Leblanc Y, Mancini J, Ouellet M, Percival D, Roy P, Skorey K, Tagari P, Vickers P, Wong E, Xu L, Prasit P (1996) From indomethacin to a selective cox-2 inhibitor: development of indolalkanoic acids as potent and selective cyclooxygenase-2 inhibitors. Bioorg Med Chem Lett 6:725–730

Copeland RA, Williams JM, Giannaras J, Nurnberg S, Covington M, Pinto D, Pick S, Trzaskos JM (1994) Mechanism of selective inhibition of the inducible isoform of prostaglandin G/H synthase. Proc Natl Acad Sci U S A 91:11202–11206

DeGray JA, Lassmann G, Curtis JF, Kennedy TA, Marnett LJ, Eling TE, Mason RP (1992) Spectral analysis of the protein-derived tyrosyl radicals from prostaglandin H synthase. J Biol Chem 267:23583–23588

DeWitt DL (1999) Cox-2-selective inhibitors: the new super aspirins. Mol Pharmacol 55:625–631

DeWitt DL, El-Harith EA, Kraemer SA, Andrews MJ, Yao EF, Armstrong RL, Smith WL (1990) The aspirin and heme-binding sites of ovine and murine prostaglandin endoperoxide synthases. J Biol Chem 265:5192–5198

Dietz R, Nastainczyk W, Ruf HH (1988) Higher oxidation states of prostaglandin H synthase. Rapid electronic spectroscopy detected two spectral intermediates during the peroxidase reaction with prostaglandin $G_2$. Eur J Biochem 171:321–328

DuBois RN, Abramson SB, Crofford L, Gupta RA, Simon LS, Van de Putte BA, Lipsky PE (1998) Cyclooxygenase in biology and disease. FASEB J 12:1063–1073

Ehrich EW, Dallob A, De Lepeleire I, Van Hecken A, Riendeau D, Yuan WY, Porras A, Wittreich J, Seibold JR, De Schepper P, Mehlisch DR, Gertz BJ (1999) Characterization of rofecoxib as a cyclooxygenase-2 isoform inhibitor and demonstration of analgesia in the dental pain model. Clin Pharmacol Ther 65:336–347

Fu J-Y, Masferrer JL, Seibert K, Raz A, Needleman P (1990) The induction and suppression of prostaglandin $H_2$ synthase (cyclooxygenase) in human monocytes. J Biol Chem 265:16737–16740

Futaki N, Takahashi S, Yokoyama M, Arai I, Higuchi S, Otomo S (1994) NS-398, a new anti-inflammatory agent, selectively inhibits prostaglandin G/H synthase/cyclooxygenase (COX-2) activity in vitro. Prostaglandins 47:55–59

Gans KR, Galbraith W, Roman RJ, Haber SB, Kerr JS, Schmidt WK, Smith C, Hewes WE, Ackerman NR (1990) Anti-inflammatory and safety profile of DuP 697, a novel orally effective prostaglandin synthesis inhibitor. J Pharmacol Exp Ther 254:180–187

Gierse JK, McDonald JJ, Hauser SD, Rangwala SH, Koboldt CM, Seibert K (1996) A single amino acid difference between cyclooxygenase-1 (COX-1) and -2 (COX-2) reverses the selectivity of COX-2 specific inhibitors. J Biol Chem 271:15810–15814

Gierse JK, Koboldt CM, Walker MC, Seibert K, Isakson PC (1999) Kinetic basis for selective inhibition of cyclooxygenases. Biochem J 339:607–614

Goodwin DC, Gunther MH, Hsi LH, Crews BC, Eling TE, Mason RP, Marnett LJ (1998) Nitric oxide trapping of tyrosyl radicals generated during prostaglandin endoperoxide synthase turnover: detection of the radical derivative of tyrosine 385. J Biol Chem 273:8903–8909

Gunther MR, Hsi LC, Curtis JF, Gierse JK, Marnett LJ, Eling TE, Mason RP (1997) Nitric oxide trapping of the tyrosyl radical of prostaglandin H synthase-2 leads to tyrosine iminoxyl radical and nitrotyrosine formation. J Biol Chem 272:17086–17090

Guo QP, Wang LH, Ruan KH, Kulmacz RJ (1996) Role of $Val^{509}$ in time-dependent inhibition of human prostaglandin H synthase-2 cyclooxygenase activity by isoform-selective agents. J Biol Chem 271:19134–19139

Hamberg M, Samuelsson B (1973) Detection and isolation of an endoperoxide intermediate in prostaglandin biosynthesis. Proc Natl Acad Sci U S A 70:899–903

Hawkey CJ (1999) COX-2 inhibitors. Lancet 353:307–314

Kalgutkar AS, Crews BC, Rowlinson SW, Garner C, Seibert K, Marnett LJ (1998) Aspirin-like molecules that covalently inactivate cyclooxgyenase-2. Science 280:1268–1270

Karthein R, Dietz R, Nastainczyk W, Ruf HH (1988) Higher oxidation states of prostaglandin H synthase. EPR study of a transient tyrosyl radical in the enzyme during the peroxidase reaction. Eur J Biochem 171:313–320

Kawamori T, Rao CV, Seibert K, Reddy BS (1998) Chemopreventive activity of celecoxib, a specific cyclooxygenase-2 inhibitor, against colon carcinogenesis. Cancer Res 58:409–412

Kujubu DA, Fletcher BS, Varnum BC, Lim RW, Herschman HR (1991) TIS10, a phorbol ester tumor promoter-inducible mRNA from Swiss 3T3 cells, encodes a novel prostaglandin synthase/cyclooxygenase homologue. J Biol Chem 266:12866–12872

Kulmacz RJ, Ren Y, Tsai A-L, Palmer G (1990) Prostaglandin H synthase: spectroscopic studies of the interaction with hydroperoxides and with indomethacin. Biochemistry 29:8760–8771

Kurumbail RG, Stevens AM, Gierse JK, McDonald JJ, Stegeman RA, Pak JY, Gildehaus D, Miyashiro JM, Penning TD, Seibert K, Isakson PC, Stallings WC (1996) Structural basis for selective inhibition of cyclooxygenase-2 by anti-inflammatory agents. Nature 384:644–648

Lambeir AM, Markey CM, Dunford HB, Marnett LJ (1985) Spectral properties of the higher oxidation states of prostaglandin H synthase. J Biol Chem 260:14894–14896

Landino LM, Crews BC, Timmons MD, Morrow JD, Marnett LJ (1996) Peroxynitrite, the coupling product of nitric oxide and superoxide, activates prostaglandin biosynthesis. Proc Natl Acad Sci U S A 93:15069–15074

Laneuville O, Breuer DK, Xu N, Huang ZH, Gage DA, Watson JT, Lagarde M, DeWitt DL, Smith WL (1995) Fatty acid substrate specificites of human prostaglandin–endoperoxide H synthase-1 and -2. Formation of 12-hydroxy-(9Z, 13E/Z, 15Z)-octadecatrienoic acids from α-linolenic acid. J Biol Chem 270:19330–19336

Lanzo CA (1998) Investigation of the binding of cyclooxygenase-selective inhibitors by fluorescence spectroscopy. Vanderbilt University, Nashville

Lecomte M, Laneuville O, Ji C, DeWitt DL, Smith WL (1994) Acetylation of human prostaglandin endoperoxide synthase-2 (cyclooxygenase-2) by aspirin. J Biol Chem 269:13207–13215

Loll PJ, Picot D, Garavito RM (1995) The structural basis of aspirin activity inferred from the crystal structure of inactivated prostaglandin $H_2$ synthase. Nat Struct Biol 2:637–642

Loll PJ, Picot D, Ekabo O, Garavito RM (1996) Synthesis and use of iodinated nonsteroidal anti-inflammatory drug analogs as crystallographic probes of the prostaglandin $H_2$ synthase cyclooxygenase active site. Biochemistry 35:7330–7340

Luong C, Miller A, Barnett J, Chow J, Ramesha C, Browner MF (1996) Flexibility of the NSAID binding site in the structure of human cyclooxygenase-2. Nat Struct Biol 3:927–933

Mancini JA, Riendeau D, Falgueyret J-P, Vickers PJ, O'Neill GP (1995) Arginine 120 of prostaglandin G/H synthase-1 is required for the inhibition by nonsteroidal anti-inflammatory drugs containing a carboxylic acid moiety. J Biol Chem 270:29372–29377

Marnett LJ, Kalgutkar AS (1998) Design of selective inhibitors of cyclooxygenase-2 as non-ulcerogenic anti-inflammatory agents. Curr Opin Chem Biol 2:482–90

Marnett LJ, Rowlinson SW, Goodwin DC, Kalgutkar AS, Lanzo CA (1999) Arachidonic acid metabolism by COX-1 and COX-2: mechanisms of catalysis and inhibition. J Biol Chem 274:22903–22906

Masferrer JL, Zweifel BS, Manning PT, Hauser SD, Leahy KM, Smith WG, Isakson PC, Seibert K (1994) Selective inhibition of inducible cyclooxygenase 2 in vivo is anti-inflammatory and nonulcerogenic. Proc Natl Acad Sci U S A 91:3228–3232

McAdam BF, Catella-Lawson F, Mardini IA, Kapoor S, Lawson JA, Fitzgerald GA (1999) Systemic biosynthesis of prostacyclin by cyclooxygenase (COX)-2: The human pharmacology of a selective inhibitor of COX-2. Proc Natl Acad Sci U S A 96:272–277

Nugteren DH, Hazelhof E (1973) Isolation and properties of intermediates in prostaglandins biosynthesis. Biochim Biophys Acta 326:448–461

O'Banion MK, Sadowski HB, Winn V, Young DA (1991) A serum- and glucocorticoid-regulated 4-kilobase mRNA encodes a cyclooxygenase-related protein. J Biol Chem 266:23261–23267

Penning TD, Talley JJ, Bertenshaw SR, Carter JS, Collins PW, Docter S, Graneto MJ, Lee LF, Malecha JW, Miyashiro JM, Rogers RS, Rogier DJ, Yu SS, Anderson GD, Burton EG, Cogburn JN, Gregory SA, Kobolt CM, Perkins WE, Siebert K, Veenhuizen AW, Zhang YY, Isakson PC (1997) Synthesis and biological evaluation of the 1,5-diarylpyrazole class of cyclooxygenase-2 inhibitors: identification of 4-[5-(4-methylphenyl)-3-(trifluoromethyl)-1H-pyrazol-1-yl]benzenesulfonamide (SC-58635, Celecoxib). J Med Chem 40:1347–1365

Picot D, Loll PJ, Garavito RM (1994) The X-ray crystal structure of the membrane protein prostaglandin $H_2$ synthase-1. Nature 367:243–249

Prasit P, Riendeau D (1997) Selective cyclooxygenase-2 inhibitors. Annu Rep Med Chem 32:211–220

Rieke CJ, Mulichak AM, Garavito RM, Smith WL (1999) The role of arginine 120 of human prostaglandin endoperoxide H synthase-2 in the interaction with fatty acid substrates and inhibitors. J Biol Chem 274:17109–17114

Rome LH, Lands WEM (1975) Structural requirements for time-dependent inhibition of prostaglandin biosynthesis by anti-inflammatory drugs. Proc Natl Acad Sci U S A 72:4863–4865

Roth GJ, Stanford N, Majerus PW (1975) Acetylation of prostaglandin synthase by aspirin. Proc Natl Acad Sci U S A 72:3073–3076

Rowlinson SW, Crews BC, Lanzo CA, Marnett LJ (1999) The Binding of arachidonic acid in the cyclooxygenase active site of mouse prostaglandin endoperoxide synthase-2 (COX-2): a putative L-shaped binding conformation utilizing the top channel region. J Biol Chem 274:23305–23310

Salvemini D, Settle SL, Masferrer JL, Seibert K, Currie MG, Needleman P (1995) Regulation of prostaglandin production by nitric oxide; an in vivo analysis. Br J Pharmacol 114:1171–1178

Shimokawa T, Kulmacz RJ, DeWitt DL, Smith WL (1990) Tyrosine 385 of prostaglandin endoperoxide synthase is required for cyclooxygenase catalysis. J Biol Chem 265:20073–20076

Simon LS, Lanza FL, Lipsky PE, Hubbard RC, Talwalker S, Schwartz BD, Isakson PC, Geis GS (1998) Preliminary study of the safety and efficacy of SC-58635, a novel cyclooxygenase 2 inhibitor: efficacy and safety in two placebo-controlled trials in osteoarthritis and rheumatoid arthritis, and studies of gastrointestinal and platelet effects. Arthritis Rheum 41:1591–1602

Singh G, Rosen-Ramey D (1998) NSAID induced gastrointestinal complications: the ARAMIS perspective – 1997. Arthritis, rheumatism, and aging medical information system. J Rheumatol Suppl 51:8–16

Smith WL, Eling TE, Kulmacz RJ, Marnett LJ, Tsai A (1992) Tyrosyl radicals and their role in hydroperoxide-dependent activation and inactivation of prostaglandin endoperoxide synthase. Biochemistry 31:3–7

Smith WL, Garavito RM, DeWitt DL (1996) Prostaglandin endoperoxide H synthases (cyclooxygenases)-1 and -2. J Biol Chem 271:33157–33160

So O-Y, Scarafia LE, Mak AY, Callan OH, Swinney DC (1998) The dynamics of prostaglandin H synthases. Studies with prostaglandin H synthase 2 Y355F unmask mechanisms of time-dependent inhibition and allosteric activation. J Biol Chem 273:5801–5807

Stubbe JS, Van der Donk WA (1998) Protein radicals in enzyme catalysis. Chem Rev 98:705–762

Takeuchi S, Yokota M, Kasai R, Ohkura Y, Tomiya T (1998) Benzoylazulene derivatives as selective cyclooxygenase-2 inhibitors. 216th ACS National Meeting. American Chemical Society, Washington, p 170

Talley JJ (1999) Selective inhibitors of cyclooxygenase-2 (COX-2). Prog Med Chem 36:201–234

Talley JJ, Bertenshaw SR, Brown DL, Carter JS, Graneto MJ, Koboldt CM, Masferrer JL, Norman BH, Rogier DJ Jr, Zweifel BS, Seibert K (1999) 4,5-Diaryloxazole inhibitors of cyclooxygenase-2 (COX-2). Med Res Rev 19:199–208

Thun MJ, Namboodiri MM, Heath CW Jr (1991) Aspirin use and reduced risk of fatal colon cancer. N Engl J Med 325:1593–1596

Tsai A-L, Palmer G, Kulmacz RJ (1992) Prostaglandin H synthase. Kinetics of tyrosyl radical formation and of cyclooxygenase catalysis. J Biol Chem 267:17753–17759

Tsai A-L, Palmer G, Xiao G, Swinney DC, Kulmacz RJ (1998) Structural characterization of arachidonyl radicals formed by prostaglandin H synthase-2 and prostaglandin H synthase-1 reconstituted with mangano protoporphyrin IX. J Biol Chem 273:3888–3894

Tsujii M, Kawano S, Tsuji S, Sawaoka H, Hori M, DuBois RN (1998) Cyclooxygenase regulates angiogenesis induced by colon cancer cells. Cell 93 (5) 705–716

Vane JR (1971) Inhibition of prostaglandin synthesis as a mechanism of action for aspirin-like drugs. Nature New Biol 231:232–235

Wallace JL (1999) Selective COX-2 inhibitors: is the water becoming muddy? Trends Pharmacol Sci 20:4–6

Warner TD, Giuliano F, Vojnovic I, Bukasa A, Mitchell JA, Vane JR (1999) Nonsteroid drug selectivities for cyclo-oxygenase-1 rather than cyclo-oxygenase-2 are associated with human gastrointestinal toxicity: A full in vitro analysis. Proc Natl Acad Sci U S A 96:7563–7568

Wong E, Bayly C, Waterman HL, Riendeau D, Mancini JA (1997) Conversion of prostaglandin G/H synthase-1 into an enzyme sensitive to PGHS-2-selective inhibitors by a double His$^{513}$→Arg and Ile$^{523}$→Val mutation. J Biol Chem 272:9280–9286

Xiao G, Tsai A-L, Palmer G, Boyar WC, Marshall PJ, Kulmacz RJ (1997) Analysis of hydroperoxide-induced tyrosyl radicals and lipoxygenase activity in aspirin-treated human prostaglandin H synthase-2. Biochemistry 36:1836–1845

Xie W, Chipman JG, Robertson DL, Erikson RL, Simmons DL (1991) Expression of a mitogen-responsive gene encoding prostaglandin synthase is regulated by mRNA splicing. Proc Natl Acad Sci U S A 88:2692–2696

Yu M, Ives D, Ramesha CS (1997) Synthesis of prostaglandin E$_2$ ethanolamide from anandamide by cyclooxygenase-2. J Biol Chem 272:21181–21186

# 5 Leukotriene-A$_4$ Hydrolase: Probing the Active Sites and Catalytic Mechanisms by Site-Directed Mutagenesis

J. Z. Haeggström and A. Wetterholm

| | | |
|---|---|---|
| 5.1 | Introduction | 85 |
| 5.2 | LTA$_4$ Hydrolase, a Zinc-Dependent Epoxide Hydrolase and Aminopeptidase | 86 |
| 5.3 | Identification of Catalytically Important Amino Acids in LTA$_4$ Hydrolase | 87 |
| 5.4 | Putative Active-Site Structure and Catalytic Mechanisms of LTA$_4$ Hydrolase | 89 |
| 5.5 | Inhibitors of LTA$_4$ Hydrolase | 92 |
| References | | 93 |

## 5.1 Introduction

Leukotrienes (LTs) are a group of bioactive lipids derived from the metabolism of polyunsaturated fatty acids, particularly arachidonic acid (Fig. 1; Samuelsson 1983). Membrane-bound arachidonic acid is released by cytosolic phospholipase A$_2$ and is further metabolized into LTA$_4$ by 5-lipoxygenase. This highly unstable epoxide may undergo enzymatic hydrolysis into the dihydroxy acid LTB$_4$ or may be conjugated with glutathione to form LTC$_4$. These two reactions are catalyzed by LTA$_4$ hydrolase and LTC$_4$ synthase, respectively. LTB$_4$ is one of the most potent chemotaxins known and has been implicated in a number of

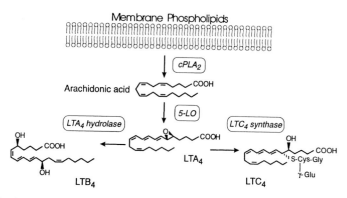

**Fig. 1.** Biosynthesis of leukotrienes (LTs). Arachidonic acid is liberated from membrane phospholipids by cytosolic phospholipase $A_2$. The fatty acid is further oxygenated and transformed into the unstable epoxide $LTA_4$. This transient intermediate is the substrate for $LTA_4$ hydrolase and $LTC_4$ synthase, which produce $LTB_4$ and $LTC_4$, respectively

acute and chronic inflammatory disorders of the skin (dermatitis), joints (arthritis), gastrointestinal tract (inflammatory bowel disease) and respiratory system (chronic obstructive pulmonary disease). This chapter describes molecular properties of $LTA_4$ hydrolase, the enzyme catalyzing the final and committed step in $LTB_4$ biosynthesis.

## 5.2 LTA₄ Hydrolase, a Zinc-Dependent Epoxide Hydrolase and Aminopeptidase

$LTA_4$ hydrolase has been purified from several mammalian sources, and complementary DNAs encoding the human, mouse, rat and guinea-pig enzymes have been cloned and sequenced (Haeggström 1997). Sequence comparisons between $LTA_4$ hydrolase and several zinc hydrolases (aminopeptidase M and thermolysin) led to the discovery of a catalytic zinc site in the enzyme (Malfroy et al. 1989; Vallee and Auld 1990). Subsequent analysis by atomic-absorption spectrometry revealed the presence of one mole of zinc per mole of protein (Haeggström et al. 1990a; Minami et al. 1990). The identification of $LTA_4$ hydrolase as a member of a family of zinc metallohydrolases, most of which are proteases or peptidases, suggested that the enzyme could possess a peptide-cleaving activity in addition to its well-characterized epoxide-hydrolase activity (the transformation of $LTA_4$ into $LTB_4$). Accordingly, $LTA_4$ hydrolase was found to hydrolyze a number of chromogenic $p$-nitroanilide or β-naphthylamide derivatives of various amino acids (Haeggström et al. 1990b; Minami et al. 1990). The physiological substrate has not yet been identified, but certain synthetic arginine tripeptides are hydrolyzed at the same rate as the lipid substrate, $LTA_4$ (Örning et al. 1994).

Both enzyme activities were inhibited by chelating compounds, e.g., 1,10-phenanthroline and 8-hydroxyquinoline-5-sulfonic acid. Metal analysis of $LTA_4$ hydrolase that had been inactivated with 1,10-phenanthroline showed that this enzyme preparation did not contain zinc and, thus, represented the apoenzyme. The primary role of the intrinsic zinc was catalytic, because the activity of the apoenzyme could be restored by the addition of stoichiometric amounts of zinc or cobalt.

The peptidase activity of $LTA_4$ hydrolase was activated by several monovalent anions (thiocyanate, chloride and bromide ions) and by albumin (Örning and Fitzpatrick 1992; Wetterholm and Haeggström 1992). The stimulatory effect of chloride obeyed saturation kinetics, suggesting the presence of an anion-binding site with an apparent affinity constant ($K_A$) of 100 mM for chloride ions. In contrast to the effect on the peptidase activity, no chloride stimulation of the epoxide hydrolase activity of the enzyme could be detected. Considering the differences in chloride concentration between the extra- and intracellular compartments, these data may be evidence of an extracellular role of the peptidase activity of $LTA_4$ hydrolase.

## 5.3 Identification of Catalytically Important Amino Acids in $LTA_4$ Hydrolase

The three proposed zinc-binding ligands, His-295, His-299 and Glu-318, were verified by site-directed mutagenesis followed by zinc analysis and activity determinations for the purified mutated proteins (Medina et al. 1991). None of the mutants contained significant amounts of zinc, and they were all enzymatically inactive.

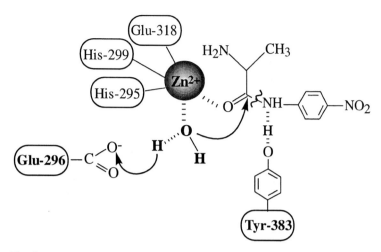

**Fig. 2.** Putative mechanism for the peptidase reaction. The fourth ligand of the catalytic zinc is an activated water molecule that is displaced by a carbonyl group of the incoming substrate. The water is polarized by the base (Glu-296) and attacks the amide bond. Simultaneously, a proton is donated from Tyr-383

Based on X-ray crystallographic studies on thermolysin, a conserved glutamic acid residue located next to the first zinc-binding ligand has been suggested to play a critical role in the reaction mechanism (Pangburn and Walsh 1975; Kester and Matthews 1977). When its counterpart in $LTA_4$ hydrolase, Glu-296, was substituted for a glutamine residue by site-directed mutagenesis, the purified mutated protein was found to be devoid of peptidase activity, while the epoxide hydrolase activity was intact or even increased compared with wild-type enzyme (Wetterholm et al. 1992). This result is in line with a role of Glu-296 as a general base in the peptidase reaction (Fig. 2).

Tyrosine 383 is another catalytically important amino acid residue in $LTA_4$ hydrolase. Sequence comparisons and results from mutational analysis suggest that this particular residue acts as a proton donor in peptidolysis (Fig. 2; Watt and Yip 1989; Minami et al. 1992; Blomster et al. 1995). Further investigation of the catalytic properties of mutants in position 383 revealed the formation of large quantities of a novel metabolite of $LTA_4$ [which was structurally identified as 5S,6S-dihydroxy-

7,9-*trans*-11,14-*cis*-eicosatetraenoic acid (5S,6S-DHETE)] in addition to the expected LTB$_4$ (Andberg et al. 1997). Hence, Tyr-383 seems to be involved in the control of the stereoselective introduction of H$_2$O at C12, because mutation of this residue shifts the specificity towards C6.

A characteristic feature of LTA$_4$ hydrolase is the inactivation and covalent modification by its substrate, LTA$_4$, which occurs during catalysis (Evans et al. 1985; Ohishi et al. 1987). Studies with electrospray mass spectrometry demonstrated a shift in molecular weight of suicide-inactivated enzyme, compatible with the binding of LTA$_4$ in a 1:1 stoichiometry between lipid and protein (Örning et al. 1992). The competitive inhibitor bestatin was able to prevent the covalent modification of the enzyme, indicating that binding occurred at the active site (Örning et al. 1992; Evans and Kargman 1992). In order to locate the region where LTA$_4$ binds to the enzyme during the inactivation process, we used differential Lys-specific peptide mapping to identify a 21-residue peptide modified by LTA$_4$ (Mueller et al. 1995b). Amino acid sequencing of this peptide, isolated from suicide-inactivated protein, demonstrated a gap in the sequence that corresponded to Tyr-378. To study the role of Tyr-378 in suicide inactivation and its potential catalytic function, we carried out a mutational analysis (Mueller et al. 1996a). Interestingly, enzyme-activity determinations before and after repeated exposure to LTA$_4$ revealed that the mutants in position 378 were no longer inactivated by LTA$_4$. Furthermore, a detailed examination of the catalytic properties revealed that mutants in position 378 were able to generate both LTB$_4$ and a second metabolite of LTA$_4$ in a yield of approximately 20–30% (Mueller et al. 1996b). This metabolite was identified as 5(S),12(R)-dihydroxy-6,10-*trans*-8,14-*cis*-eicosatetraenoic acid, i.e., $\Delta^6$-*trans*-$\Delta^8$-*cis*-LTB$_4$.

## 5.4 Putative Active-Site Structure and Catalytic Mechanisms of LTA$_4$ Hydrolase

If one compiles information that has been generated from biochemical studies, sequence comparisons and mutational analysis, a model of the active site of LTA$_4$ hydrolase may look as outlined in Fig. 3. It appears that the two enzyme activities are exerted via overlapping active sites and, therefore, one can envisage two closely related substrate-binding

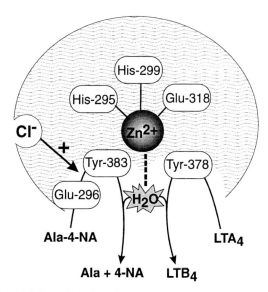

**Fig. 3.** Model of the active site of leukotriene-$A_4$ hydrolase. The two active sites corresponding to the epoxide hydrolase and peptidase reactions overlap. The catalytic zinc, complexed to His-295, His-299 and Glu-318, is near both pockets, with the water at attacking distance from the substrate(s). Chloride specifically stimulates the peptidase activity. Three active-site residues (Glu-296 and Tyr-383, involved in the peptidase reaction, and Tyr-378, a structural determinant for suicide inactivation) are indicated

pockets, one for a peptide substrate not yet identified and the other for the lipid substrate $LTA_4$. Furthermore, results with chemical modification have indicated that a basic residue acts as a carboxylate-recognition site and contributes to substrate binding (Mueller et al. 1995a). Since both the epoxide hydrolase and peptidase activities are zinc-dependent, the metal should be located at (or at least close to) the junction of the substrate-binding pockets.

Regarding the peptidase activity, a catalytic model based on data for thermolysin may be proposed (Pangburn and Walsh 1975; Kester and Matthews 1977). Thus, the catalytic zinc is complexed to His-295, His-299, Glu-318 and an activated water molecule. The water is displaced from the zinc atom by the carbonyl oxygen of the substrate and is then polarized by the carboxylate of Glu-296 to promote an attack on the

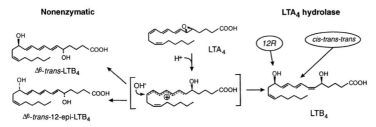

**Fig. 4.** Enzymatic and nonenzymatic hydrolysis of leukotriene A4 (LTA$_4$). In both the enzymatic and non-enzymatic hydrolysis of LTA$_4$, a carbocation is formed according to an S$_N$1 mechanism. Water may be added to this intermediate non-enzymatically via attack from either side of the molecule, which generates the epimers (at C12) of all-*trans*-LTB$_4$. Alternatively, the epoxide may be enzymatically hydrolyzed by LTA$_4$ hydrolase, with formation of a hydroxyl group in the R configuration and a conjugated triene in the *cis–trans–trans* configuration

carbonyl carbon of the scissile peptide bond. At the same time, a proton is transferred to the nitrogen of the peptide bond by Tyr-383 (Fig. 2).

Much less is known about amino acid residues involved in the conversion of LTA$_4$ into LTB$_4$. If one considers non-enzymatic hydrolysis of LTA$_4$, this process is thought to proceed according to an S$_N$1 mechanism. According to that model, the reaction is initiated via an acid-induced opening of the epoxide moiety and the formation of a carbonium ion, with a positive charge delocalized over the triene structure (Fig. 4). This intermediate has a planar $sp^2$-hybridized configuration at C12, which in turn allows a nucleophilic attack from both sides of the carbon. Accordingly, the two epimers at C12 of 5(S),12-dihydroxy-6,8,10-*trans*-14-*cis*-eicosatetraenoic acid are formed. The structure of LTB$_4$ (formed by enzymatic hydrolysis) differs from the structure of either of the two non-enzymatically formed 5,12-dihydroxy acids in two ways: the double-bond geometry and the configuration of the hydroxyl group at C12. Apparently, the role of LTA$_4$ hydrolase during enzymatic hydrolysis of LTA$_4$ into LTB$_4$ is to ensure a stereoselective introduction of H$_2$O at C12 to yield the 12-R epimer of the hydroxyl group and to form the $\Delta^6$-*cis*-$\Delta^8$-*trans*-$\Delta^{10}$-*trans* configuration of the conjugated triene. Interestingly, the mutants at position 378 of LTA$_4$ hydrolase differ from wild-type enzyme in one of the two essential functions of the enzyme, i.e., the positioning of the *cis* double bond in the product. Hence, Tyr-378 appears to be involved in this aspect of catalysis, perhaps by assisting in the proper alignment of $LTA_4$ in the substrate-binding pocket or by promoting a favorable conformation of a putative carbonium-ion intermediate. Considering that Tyr-378 is an important structural determinant of suicide inactivation, this residue should be positioned in the vicinity of the reactive allylic epoxide and the conjugated triene of $LTA_4$.

An additional clue to the mechanism of $LTB_4$ biosynthesis comes from the finding that mutants of Tyr-383, as previously mentioned, generate substantial amounts of 5S,6S-DHETE in addition to $LTB_4$. The stereochemistry of this compound implies an $S_N1$ mechanism in its formation, which in turn indicates that conversion of $LTA_4$ into $LTB_4$ proceeds according to the same mechanism. This model agrees well with a recent study in which the active site of $LTA_4$ hydrolase was probed by a series of structural analogs of amino hydroxamic acids, potent, tight-binding inhibitors of the enzyme. The results of this work indicated that the zinc acts as a Lewis acid and activates the epoxide to initiate an $S_N1$-type reaction.

## 5.5 Inhibitors of $LTA_4$ Hydrolase

The identification of $LTA_4$ hydrolase as a zinc metalloenzyme with a peptide-cleaving activity suggested that the enzyme could be sensitive to known inhibitors of zinc peptidases. Thus, bestatin and captopril were found to be effective inhibitors of $LTA_4$ hydrolase (Örning et al. 1991a, 1991b). Several laboratories have developed more powerful and selective inhibitors. For instance, an α-keto-β-amino ester and a thioamine were found to be potent, tight-binding inhibitors with $IC_{50}$ values in the low-micromolar to nanomolar range (Yuan et al. 1991, 1992, 1993; Wetterholm et al. 1995). In addition, a series of β-amino-hydroxylamine and amino-hydroxamic acids, some of which were equipotent to the above mentioned substances, were synthesized (Hogg et al. 1995). Kelatorphan (a known inhibitor of enkephalin-degrading enzymes) and several related analogs were also found to be potent inhibitors of both enzyme activities of $LTA_4$ hydrolase (Penning et al. 1995). Perhaps the most potent compound described thus far is SC-57461 (*N*-methyl-*N*-[3-

[4-(phenylmethyl)-phenoxy]propyl]-β-alanine; Yuan et al. 1996), which was orally active and showed very promising results in an animal model of colitis (Smith et al. 1996). Interestingly, it has also been reported that inhibitors of $LTA_4$ hydrolase can act in synergy with inhibitors of cyclo-oxygenase, as shown for the compounds SA-6541 and indomethacin in carrageenan-induced murine dermatitis (Tsuji et al. 1998).

**Acknowledgments.** We are grateful to Juan F. Medina, Martin J. Mueller, Martina Blomster, Filippa Kull, Eva Ohlson and Bengt Samuelsson for their contributions to the studies described in this chapter. Work in the author's laboratory was financially supported by the Swedish Medical Research Council (O3X-10350) and Konung Gustav V;s 80-årsfond.

## References

Andberg MB, Hamberg M, Haeggström JZ (1997) Mutation of tyrosine 383 in leukotriene $A_4$ hydrolase allows formation of 5S,6S-dihydroxy-7,9-*trans*-11,14-*cis*-eicosatetraenoic acid: Implications for the epoxide hydrolase mechanism. J Biol Chem 272:23057–23063

Blomster M, Wetterholm A, Mueller MJ, Haeggström JZ (1995) Evidence for a catalytic role of tyrosine 383 in the peptidase reaction of leukotriene $A_4$ hydrolase. Eur J Biochem 231:528–534

Evans JF, Kargman S (1992) Bestatin inhibits covalent coupling of [$^3$H]$LTA_4$ to human leukocyte $LTA_4$ hydrolase. FEBS Lett 297:139–142

Evans JF, Nathaniel DJ, Zamboni RJ, Ford-Hutchinson AW (1985) Leukotriene $A_3$: a poor substrate but a potent inhibitor of rat and human neutrophil leukotriene $A_4$ hydrolase. J Biol Chem 260:10966–10970

Haeggström JZ (1997) The molecular biology of the leukotriene $A_4$ hydrolase. In: Dahlén SE, Holgate S (eds) From SRS-A to leukotrienes. Blackwell, Oxford, p 85

Haeggström JZ, Wetterholm A, Shapiro R, Vallee BL, Samuelsson B (1990a) Leukotriene $A_4$ hydrolase: a zinc metalloenzyme. Biochem Biophys Res Commun 172:965–970

Haeggström JZ, Wetterholm A, Vallee BL, Samuelsson B (1990b) Leukotriene $A_4$ hydrolase: an epoxide hydrolase with peptidase activity. Biochem Biophys Res Commun 173:431–437

Hogg JH, Ollmann IR, Haeggström JZ, Wetterholm A, Samuelsson B, Wong C-H (1995) Amino hydroxamic acids as potent inhibitors of $LTA_4$ hydrolase. Bioorg Med Chem 3:1405–1415

Kester WR, Matthews BW (1977) Crystallographic study of the binding of dipeptide inhibitors to thermolysin: Implications for the mechanism of catalysis. Biochemistry 16:2506–2516

Malfroy B, Kado-Fong H, Gros C, Giros B, Schwartz J-C, Hellmiss R (1989) Molecular cloning and amino acid sequence of rat kidney aminopeptidase M: a member of a super family of zinc-metallohydrolases. Biochem Biophys Res Commun 161:236–241

Medina JF, Wetterholm A, Rådmark O, Shapiro R, Haeggström JZ, et al. (1991) Leukotriene $A_4$ hydrolase: determination of the three zinc-binding ligands by site directed mutagenesis and zinc analysis. Proc Natl Acad Sci U S A 88:7620–7624

Minami M, Ohishi N, Mutoh H, Izumi T, Bito H, et al. (1990) Leukotriene $A_4$ hydrolase is a zinc-containing aminopeptidase. Biochem Biophys Res Commun 173:620–626

Minami M, Bito H, Ohishi N, Tsuge H, Miyano M, et al. (1992) Leukotriene $A_4$ hydrolase, a bifunctional enzyme. Distinction of leukotriene $A_4$ hydrolase and aminopeptidase activities by site-directed mutagenesis at Glu-297. FEBS Lett 309:353–357

Mueller MJ, Samuelsson B, Haeggström JZ (1995a) Chemical modification of leukotriene A4 hydrolase. Indications for essential tyrosyl and arginyl residues at the active site. Biochemistry 34:3536–3543

Mueller MJ, Wetterholm A, Blomster M, Jörnvall H, Samuelsson B, Haeggström JZ (1995b) Leukotriene $A_4$ hydrolase: mapping of a heneicosapeptide involved in mechanism-based inactivation. Proc Natl Acad Sci U S A 92:8383–8387

Mueller MJ, Blomster M, Opperman UCT, Jörnvall H, Samuelsson B, Haeggström JZ (1996a) Leukotriene $A_4$ hydrolase: protection from mechanism-based inactivation by mutation of tyrosine-378. Proc Natl Acad Sci U S A 93:5931–5935

Mueller MJ, Blomster M, Samuelsson B, Haeggström JZ (1996b) Leukotriene $A_4$ hydrolase: mutation of tyrosine-383 allows conversion of leukotriene $A_4$ into an isomer of leukotriene $B_4$. J Biol Chem 271:24345–24348

Ohishi N, Izumi T, Minami M, Kitamura S, Seyama Y, et al. (1987) Leukotriene $A_4$ hydrolase in the human lung: inactivation of the enzyme with leukotriene $A_4$ isomers. J Biol Chem 262:10200–10205

Örning L, Fitzpatrick FA (1992) Albumins activate peptide hydrolysis by the bifunctional enzyme $LTA_4$ hydrolase/aminopeptidase. Biochemistry 31:4218–4223

Örning L, Krivi G, Bild G, Gierse J, Aykent S, Fitzpatrick FA (1991a) Inhibition of leukotriene $A_4$ hydrolase/aminopeptidase by captopril. J Biol Chem 266:16507–16511

Örning L, Krivi G, Fitzpatrick FA (1991b) Leukotriene $A_4$ hydrolase: Inhibition by bestatin and intrinsic aminopeptidase activity establish its functional resemblance to metallohydrolase enzymes. J Biol Chem 266:1375–1378

Örning L, Gierse J, Duffin K, Bild G, Krivi G, Fitzpatrick FA (1992) Mechanism-based inactivation of leukotriene $A_4$ hydrolase/aminopeptidase by leukotriene $A_4$. Mass spectrometric and kinetic characterization. J Biol Chem 267:22733–22739

Örning L, Gierse JK, Fitzpatrick FA (1994) The bifunctional enzyme leukotriene-$A_4$ hydrolase is an arginine aminopeptidase of high efficiency and specificity. J Biol Chem 269:11269–11273

Pangburn MK, Walsh KA (1975) Thermolysin and neutral protease: mechanistic considerations. Biochemistry 14:4050–4054

Penning TD, Askonas LJ, Djuric SW, Haack RA, Yu SS, et al. (1995) Kelatorphan and related analogs – potent and selective inhibitors of leukotriene $A_4$ hydrolase. Bioorg Med Chem 5:2517–2522

Samuelsson B (1983) Leukotrienes: mediators of immediate hypersensitivity reactions and inflammation. Science 220:568–575

Smith WG, Russell MA, Liang C, Askonas LA, Kachur JF, et al. (1996) Pharmacological characterization of selective inhibitors of leukotriene $A_4$ hydrolase. Prostaglandins Leukot Essent Fatty Acids 55[suppl]1:13

Tsuji F, Miyake Y, Enomoto H, Horiuchi M, Mita S (1998) Effects of SA6541, a leukotriene $A_4$ hydrolase inhibitor, and indomethacin on carrageenan-induced murine dermatitis. Eur J Pharmacol 346:81–85

Vallee BL, Auld DS (1990) Zinc coordination, function, and structure of zinc enzymes and other proteins. Biochemistry 29:5647–5659

Watt VM, Yip CC (1989) Amino acid sequence deduced from rat kidney cDNA suggests it encodes the Zn-peptidase aminopeptidase N. J Biol Chem 264:5480–5487

Wetterholm A, Haeggström JZ (1992) Leukotriene $A_4$ hydrolase: an anion activated peptidase. Biochim Biophys Acta 1123:275–281

Wetterholm A, Medina JF, Rådmark O, Shapiro R, Haeggström JZ, et al. (1992) Leukotriene $A_4$ hydrolase: abrogation of the peptidase activity by mutation of glutamic acid-296. Proc Natl Acad Sci U S A 89:9141–9145

Wetterholm A, Haeggström JZ, Samuelsson B, Yuan W, Munoz B, Wong C (1995) Potent and selective inhibitors of leukotriene $A_4$ hydrolase: effects on purified enzyme and human polymorphonuclear leukocytes. J Pharmacol Exp Ther 275:31–37

Yuan W, Zhong Z, Wong CH, Haeggström JZ, Wetterholm A, Samuelsson B (1991) Probing the inhibition of leukotriene $A_4$ hydrolase based on its aminopeptidase activity. Bioorg Med Chem 1:551–556

Yuan W, Wong C-H, Haeggström JZ, Wetterholm A, Samuelsson B (1992) Novel tight-binding inhibitors of leukotriene A4 hydrolase. J Am Chem Soc 114:6552–6553

Yuan W, Munoz B, Wong C-H, Haeggström JZ, Wetterholm A, Samuelsson B (1993) Development of selective tight-binding inhibitors of leukotriene $A_4$ hydrolase. J Med Chem 36:211–220

Yuan JH, Birkmeier J, Yang DC, Hribar JD, Liu N, et al. (1996) Isolation and identification of metabolites of leukotriene $A_4$ hydrolase inhibitor SC-57461 in rats. Drug Metab Dispos 24:1124–1133

# 6 The Regulation of Cyclooxygenase-1 and -2 in Knockout Cells and Cyclooxygenase and Fever in Knockout Mice

L. R. Ballou

| | | |
|---|---|---|
| 6.1 | Overview of Findings | 97 |
| 6.2 | Introduction | 99 |
| 6.3 | Results and Discussion | 102 |
| 6.4 | Concluding Remarks | 120 |
| References | | 121 |

## 6.1 Overview of Findings

### 6.1.1 Regulation Of Cyclooxygenase-1 and -2 in Knockout Cells

Prostaglandin $E_2$ ($PGE_2$) production in immortalized, non-transformed lung cells derived from wild-type, cyclooxygenase-1 (COX-1)$^{-/-}$ or COX-2$^{-/-}$ deficient mice was examined after treatment with interleukin-1β (IL-1), tumor necrosis factor α (TNF), acidic fibroblast growth factor (FGF) and phorbol ester (PMA). Compared with their wild-type (COX-1$^{+/+}$/COX-2$^{+/+}$) counterparts, COX-1$^{-/-}$ or COX-2$^{-/-}$ cells exhibited substantially enhanced expression of the remaining functional COX gene. Furthermore, both basal and IL-1-induced expression of cytosolic phospholipase $A_2$ ($cPLA_2$), a key enzyme regulating substrate mobilization for $PGE_2$ biosynthesis, was more pronounced in both COX-1- and COX-2-deficient cells. Thus, COX-1$^{-/-}$ and COX-2$^{-/-}$ cells have the ability to co-ordinate the upregulation of the alternate COX isoenzyme and $cPLA_2$ genes to overcome defects in prostaglandin (PG)-biosynthetic machinery. The potential for cells to alter (and thereby compensate for) defects in the expression of specific genes (such as COX) has significant clinical implications, given the central role of COX in pathophysiology and the widespread use of COX inhibitors as therapeutic agents (Kirtikara et al. 1998).

### 6.1.2 COX and Fever in Knockout Mice

Many lines of evidence have implicated inducible cyclooxygenase-2 (COX-2) in fever production. Most notably, COX-2 expression is selectively enhanced in brain tissue after peripheral exogenous (lipopolysaccharide; LPS) or endogenous (IL-1) pyrogen administration, while selective COX-2 inhibitors suppress the fever induced by these pyrogens. Therefore, we assessed the febrile response to LPS in wild-type, COX-$1^{-/-}$ and COX-$2^{-/-}$ mice, which were pre-trained with the experimental conditions daily for 2 weeks. LPS was injected intraperitoneally at 1 µg/mouse; pyrogen-free saline (PFS) was the vehicle and control solution. Core temperatures ($T_c$s) were recorded using thermocouples inserted 2 cm into the colon. The presence of the COX isoforms was determined immunocytochemically in the cerebral blood vessels after the experiments, without knowledge of the functional results. Wild-type, COX-$1^{+/-}$ (heterozygous) and COX-$1^{-/-}$ (homozygous) mice all responded to LPS with a 1°C rise in $T_c$ within 1 h; the fever gradually abated over the next 4 h. In contrast, COX-$2^{+/-}$ and COX-$2^{-/-}$ mice displayed no $T_c$ rise after LPS. PFS did not affect the $T_c$ of any animal. Thus, it appears that COX-2 is necessary for LPS-induced fever production (Li et al. 1999).

## 6.2 Introduction

### 6.2.1 Regulation of PG Biosynthesis

PGs, such as $PGE_2$, are pivotal modulators of tissue homeostasis, and their aberrant regulation is known to cause serious pathophysiological consequences (Goetzl et al. 1995; Herschman et al. 1995; Herschman 1996; Smith et al. 1996). $PGE_2$ biosynthesis is regulated by successive metabolic steps involving the phospholipase $A_2$ ($PLA_2$)-mediated release of arachidonic acid (AA) and its conversion to $PGE_2$ by COX, hydroperoxidase and isomerase activities (Goetzl et al. 1995; Herschman et al. 1995; Herschman 1996; Smith et al. 1996). Although cytosolic $PLA_2$ ($cPLA_2$) is primarily responsible for agonist-induced AA release from membrane phospholipids (Clark et al. 1991; Lin et al. 1992), secretory $PLA_2$ ($sPLA_2$) may also be important in regulating AA availability via a transcellular mechanism (Reddy and Herschman 1996). Conversion of AA to $PGH_2$, the committed step in prostanoid biosynthesis, is mediated by COX-1 and COX-2, which are encoded by two unique genes located on different chromosomes (Smith et al. 1996). Generally, while COX-1 is constitutively expressed, the expression of COX-2 is highly inducible (Herschman 1996; Smith et al. 1996). Based on their respective modes of expression, it is thought that COX-1 is primarily involved in cellular homeostasis, while COX-2 plays a major role in inflammation and mitogenesis. The COX isoenzymes are the primary targets for non-steroidal anti-inflammatory drugs (NSAIDs), which act by inhibiting the COX activity of COX-1 and COX-2, thereby blocking their ability to convert AA to $PGG_2$ (DeWitt et al. 1993; Vane and Botting 1995b). In addition to their use as analgesics and for alleviation of acute and chronic inflammation, NSAIDs have proven effective in decreasing the frequency of heart attacks and strokes (Vane and Botting 1994, 1995, 1995) and in reducing the incidence of colon cancer (Rigas et al. 1993; Kargman et al. 1995).

Since most cells are capable of expressing both COX-1 and COX-2, it has been difficult to dissect the precise roles of the COX isoenzymes with respect to their relative contributions to the myriad biochemical events involved in eliciting immune and inflammatory responses. The purpose of the present study was to examine the effects of COX deficiency on the expression of the COX-1 and COX-2 isoenzymes and to compare the responses of wild-type, COX-1 and COX-2 knockout cells with respect to agonist-induced $PGE_2$ biosynthesis. We demonstrated that the expression of alternative COX isoforms and $cPLA_2$ and $PGE_2$ production are significantly increased in COX-deficient cells. Thus, COX deficiency, regardless of whether it is COX-1 or COX-2, results in enhanced basal and inducible expression of the remaining COX isoenzyme and the elevated expression of $cPLA_2$. We interpret these data to indicate that the elevated production of $PGE_2$ in COX-1 or -2 isoenzyme-deficient cells is due to the compensatory expression of the remaining COX isoenzyme (Kirtikara et al. 1998). Based on these findings, we are now determining whether COX "compensation" is an in vivo phenomenon in mice congenitally lacking one of the COX isoenzymes. We are also examining the respective roles of COX-1 and COX-2 in COX knockout mice using a number of models for acute (air pouch) and immune-mediated inflammatory processes, such as collagen-induced arthritis (CIA; Rosloneic 1999).

### 6.2.2 COX and Fever

Much evidence supports the role of $PGE_2$ as the proximal mediator of fever (Coceani 1991; Blatteis and Schic 1997). $PGE_2$ is produced and released in response to pyrogenic cytokines (endogenous pyrogens; EnPs), which are themselves produced and released in response to invading infectious pathogens or their products, exogenous pyrogens (ExPs). $PGE_2$ is implicated as a fever mediator because:

1. It is a potent hyperthermic agent (Coceani 1991; Blatteis and Schic 1997) thought to act directly or indirectly on thermoregulatory neurons in the preoptic anterior hypothalamus, the primary brain site in which body temperature is regulated (Boulant 1996)
2. Its level increases and decreases in the brain region in conjunction with the febrility course (Coceani et al. 1983; Sehic et al. 1996)
3. COX inhibitors (NSAIDs), inhibit pyrogen fever in parallel with the inhibition of $PGE_2$ synthesis (Milton 1990)
4. The congenital absence of the $PGE_2$ receptor $EP_3$ impairs the febrile responses to ExP and EnP (Ushikubi et al. 1998).

Although $PGE_2$ levels in blood rise promptly after the entry of micro-organisms or after systemic ExP or EnP administration (Rotondo et al. 1988; Milton 1990), it is now generally agreed that the $PGE_2$ detected in the brain is not derived from the blood but is instead produced directly in the brain (Coceani et al. 1988; Morimoto et al. 1992), although some species differences may exist (Eguchi et al. 1988). However, the precise cell source and nature of the triggering mechanism that induces $PGE_2$ in the brain in response to systemic pyrogens and its precise mode of action remain controversial.

Because it can be induced by inflammatory mediators, it is likely that COX-2 plays a more significant role in the brain in fever production than does COX-1. Indeed, it is now well documented that ExPs (LPS) and EnPs (IL-1) activate COX-2 expression in vivo (Chan et al. 1995; Cao et al. 1997, 1998); however, it appears to be constitutively expressed in unstimulated neurons (Breder et al. 1992, 1995; O'Neill and Ford-Hutchinson 1993; Yamagata et al. 1993; Kaufmann et al. 1996; Quan et al. 1998) and may or may not be upregulated by pyrogenic stimuli (Coceani 1991; Breder 1997). In contrast, COX-1 expression was not affected by the peripheral administration of pyrogens anywhere in the brain. Moreover, treatment with specific COX-2 inhibitors [NS-398, L-745, L-337, 5,5-dimethyl-3-(3-fluorophenyl)-4-(4-methylsulphonyl)phenyl-2-(5-H)-furanone] orally after i.v. LPS (Futaki et al. 1994) or intraperitoneally before intraperitoneal (i.p.) LPS (Chan et al. 1995; Cao et al. 1998) suppressed the febrile response but did not affect basal body temperature. These antipyretic effects were not different from those produced by conventional NSAIDs, which inhibit both COX-1 and COX-2. Taken together, therefore, these data seem to provide compelling support for the critical importance of COX-2 in fever genesis. To substantiate this inference, we examined the febrile response to LPS administered intraperitoneally to COX-1 and COX-2 knockout mice (Li et al. 1999).

## 6.3 Results and Discussion

### 6.3.1 Regulation of COX-1 and COX-2 in Knockout Cells

#### 6.3.1.1 The Effect of IL-1 on $PGE_2$ Biosynthesis, COX-1 and COX-2 Expression

We examined the effects of IL-1 on $PGE_2$ production in cells containing both COX isoenzymes (wild-type) compared with cells which had only COX-1 ($COX-2^{-/-}$) or COX-2 ($COX-1^{-/-}$), respectively. As shown in Fig. 1A, $COX-1^{-/-}$ or $COX-2^{-/-}$ cells synthesized six- to eightfold higher amounts of $PGE_2$ compared with their wild-type counterparts. Interestingly, basal $PGE_2$ production was higher in both $COX-2^{-/-}$ (66.71±3.54 pg/$10^3$ cells; $n=6$) and $COX-1^{-/-}$ (90.23±3.29 pg/$10^3$ cells; $n=8$) cells compared with wild-type [11.07±0.62 pg/$10^3$ cells; $n=8$, all values are means±standard errors (SE)]. IL-1 treatment of wild-type and $COX-1^{-/-}$ cells further enhanced their $PGE_2$ output. In contrast, IL-1 treatment of $COX-2^{-/-}$ cells did not significantly enhance $PGE_2$ biosynthesis.

The dramatic differences in basal $PGE_2$ biosynthesis between wild-type and COX-deficient cells prompted us to compare the expression of genes encoding three key enzymes (COX-1, COX-2 and $cPLA_2$) regulating $PGE_2$ biosynthesis in untreated and IL-1-treated wild-type, $COX-1^{-/-}$ and $COX-2^{-/-}$ cells. A comparison of basal and IL-1-stimulated levels of COX-1 and COX-2 proteins via immunoblot assay in wild-type and $COX-1^{-/-}$ cells is shown in Figs. 1B and 1C. Consistent with numerous previous observations, the basal expression of COX-2 protein in wild-type cells was barely detectable. Constitutive levels of COX-2 proteins were also significantly increased (~2.4-fold) in untreated $COX-1^{-/-}$ cells. The elevated level of COX-2 protein correlates well with the higher basal $PGE_2$ levels in $COX-1^{-/-}$ cells compared with those in wild-type cells. When treated with IL-1, COX-2 protein levels increased moderately in wild-type cells, but the increase in COX-2 protein was much more dramatic in $COX-1^{-/-}$ cells (~40-fold). The overall pattern of COX-2 protein expression in wild-type and $COX-1^{-/-}$ cells correlated with increased $PGE_2$ production seen in cells with unique COX phenotypes (Fig. 1A).

Next, we examined the basal and IL-1-stimulated levels of COX-1 protein in identically treated wild-type, $COX-2^{-/-}$ and $COX-1^{-/-}$ cells,

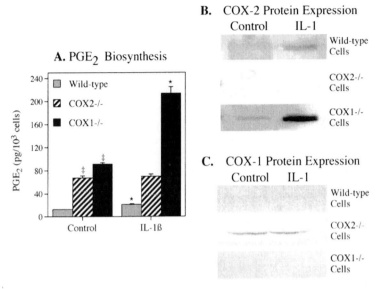

**Fig. 1.** The effect of interleukin 1 (*IL-1*) on prostaglandin $E_2$ (*$PGE_2$*) biosynthesis cyclooxygenase 1 (*COX-1*) and COX-2 expression. **A** Wild-type, COX-$2^{-/-}$ and COX-$1^{-/-}$ cells were treated with vehicle (control) or with IL-1 (0.25 ng/ml). After 24 h, culture media was collected and analyzed for $PGE_2$ by radioimmunoassay. Data are means ± standard errors of at least six separate determinations (wells). ‡ Indicates values significantly different from the wild-type value in the control group (as determined with a paired *t*-test; $P<0.05$). *Asterisks* indicate values significantly different from the control values of each respective cell type (as determined with a paired *t*-test). **B** Western-blot analysis of COX-2 protein levels in IL-1-stimulated cells. Wild-type, COX-$2^{-/-}$ and COX-$1^{-/-}$ cells were treated with vehicle (control) or with IL-1 (0.25 ng/ml). After 24 h, culture media was removed, and Western-blot analysis was performed. **C** Western-blot analysis of COX-1 protein levels in IL-1-stimulated cells. Wild-type, COX-$2^{-/-}$ and COX-$1^{-/-}$ cells were treated with vehicle (control) or with IL-1 (0.25 ng/ml). After 24 h, culture media was removed and Western-blot analysis was performed (Kirtikara et al. 1998)

respectively (Fig. 1C). In wild-type cell extracts, the level of COX-1 protein was barely detectable, and IL-1 treatment was apparently inconsequential. This result was not unexpected, because COX-1 expression is not known to be inducible under many conditions. We observed that basal expression of COX-1 protein in untreated COX-2$^{-/-}$ cells was much greater (~14-fold) than that in wild-type cells. This overexpression of COX-1 protein corresponds to greater basal PGE$_2$ levels in COX-2$^{-/-}$ cells compared with the basal levels in wild-type cells. IL-1 had no stimulatory effect on COX-1 protein levels in COX-2$^{-/-}$ cells and, as expected, COX-1$^{-/-}$ cells did not express detectable COX-1 protein. Another important enzyme in the PG-biosynthesis pathway is PGE$_2$ synthase, the isomerase that converts PGH$_2$ to PGE$_2$. Although PGE$_2$ synthase has neither been sequenced nor cloned, making it difficult to study, available evidence does seem to indicate that this enzyme is not a rate-limiting reagent in PGE$_2$ biosynthesis. However, based on our findings, we cannot rule out the possibility that PGE$_2$-synthase expression may be altered in COX null cells.

In order to examine the possibility that iso-PGE$_2$ or other isoprostanes (Roberts and Morrow 1997) may be generated non-enzymatically from a build-up of endoperoxide intermediates that cross-react with the anti-PGE$_2$ used in our radioimmunoassay, leading to erroneously high estimations of COX and/or PGE$_2$ synthase activity, we performed two experiments. First, we treated wild-type and COX$^{-/-}$ cells with either indomethacin or NS-398 (COX-1- and COX-2-selective inhibitors, respectively), because these COX inhibitors should block PGE$_2$ synthesis without affecting iso-PGE$_2$ formation. We found that either indomethacin or NS-398 completely block both the basal and cytokine-induced formation of immunoreactive PGE$_2$ in wild-type and COX$^{-/-}$ cells (data not shown). Second, radio thin-layer chromatography was used to confirm that PGE$_2$ was the predominant prostanoid product generated by wild-type and COX$^{-/-}$ cells and that no other AA metabolites aside from PGE$_2$ were generated in COX$^{-/-}$ cells (data not shown).

### 6.3.1.2 *The Comparative Effects of Cytokines and PMA on PGE$_2$ Biosynthesis*

In order to compare the effects of IL-1 (Fig. 1A) to those of other inducers of PGE$_2$ biosynthesis, we tested the effects of TNF, acidic FGF, and PMA on PGE$_2$ production in wild-type, COX-2$^{-/-}$ and COX-1$^{-/-}$

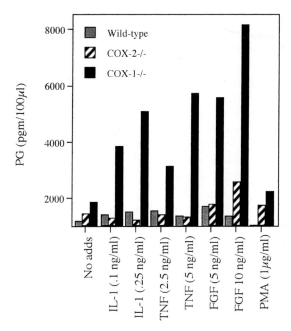

**Fig. 2.** The comparative effects of cytokines and phorbol myristate acetate (PMA) on prostaglandin $E_2$ ($PGE_2$) biosynthesis. Wild-type, cyclooxygenase 2 (COX-2)$^{-/-}$ and COX-1$^{-/-}$ mouse lung cells were treated with interleukin 1 (0.25 ng/ml), tumor necrosis factor (5 ng/ml), fibroblast growth factor (10 ng/ml) or PMA (12.5 ng/ml). After 24 h, culture media was collected and analyzed for $PGE_2$ by radioimmunoassay. Data are means ± standard errors of at least six separate determinations (wells). *Asterisks* indicate values significantly different from the control values of each respective cell type (as determined with a paired *t*-test; $P<0.05$; Kirtikara et al. 1998)

cells. Compared with stimulated wild-type cells, there was significantly more $PGE_2$ produced in either COX-1$^{-/-}$ or COX-2$^{-/-}$ cells (with the possible exception of TNF, which induced comparable $PGE_2$ biosynthesis in each cell type). In response to FGF, the amount of $PGE_2$ produced by COX-2$^{-/-}$ cells was elevated, and $PGE_2$ was even more dramatically elevated in COX-1$^{-/-}$ cells compared with wild-type cells. COX-2$^{-/-}$ and COX-1$^{-/-}$ cells treated with PMA also produced much more $PGE_2$ than did wild-type cells (Fig. 2). Thus, in general, COX-isoenzyme deficiency results in increased $PGE_2$ biosynthesis, but the relative contributions of COX-1 and COX-2 are clearly dependent on the specific agonists involved.

#### 6.3.1.3 Elevated cPLA$_2$ Expression in COX Knockout Cells

Since constitutive COX-2 protein expression and $PGE_2$ production in COX-1$^{-/-}$ cells was significantly enhanced, we were also curious about the status of $cPLA_2$ gene expression in COX-deficient cells. We reasoned that $cPLA_2$ activity could be involved in regulating levels of free AA for conversion into $PGE_2$; thus, $cPLA_2$ could play a critical role in compensating for COX-isoenzyme deficiency. We were somewhat surprised to find that basal levels of $cPLA_2$ protein in COX-2$^{-/-}$ and COX-1$^{-/-}$ cells were significantly higher than levels of $cPLA_2$ in wild-type cells (Fig. 3). It is conceivable, therefore, that enhanced expression of $cPLA_2$ could directly contribute to higher $PGE_2$ levels in both COX-deficient cells by generating more AA substrates for $PGE_2$ biosynthesis. Treatment of COX-1$^{-/-}$ or COX-2$^{-/-}$ cells with IL-1 resulted in a modest increase in the amount of $cPLA_2$ protein (an approximately fourfold increase in COX-1$^{-/-}$ and an approximately 1.4-fold increase in COX-2$^{-/-}$). This was in contrast to wild-type cells, which showed no change in the levels of $cPLA_2$ protein after treatment with IL-1. As an important control, we examined the quantitative parameters of $PGE_2$ production and COX-1, COX-2 and $cPLA_2$ gene expression in wild-type, COX-2$^{-/-}$ and COX-1$^{-/-}$ cells from primary cell cultures and found essentially the same patterns in primary cells as were observed in the immortalized cells (data not shown). Therefore, the characteristic pattern of expression of COX-1, COX-2 and $cPLA_2$ proteins in COX-2$^{-/-}$, and COX-1$^{-/-}$ cells is not elicited as a result of immortalization caused by the E1A

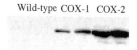

**Fig. 3.** Elevated cytosolic-phospholipase-A$_2$ expression in cyclooxygenase (*COX*) knockout cells. Wild-type, COX-2$^{-/-}$ and COX-1$^{-/-}$ mouse cells were treated with vehicle (control) or with interleukin 1 (0.25 ng/ml). After 24 h, culture media was removed, and Western-blot analysis was performed. **A** Protein from wild-type cells. **B** Protein from COX-2$^{-/-}$ cells. **C** Protein from COX-1$^{-/-}$ cells (Kirtikara et al. 1998)

adenovirus gene. Taken together, these data indicate that COX-1$^{-/-}$ cells express enhanced levels of both basal and cytokine-stimulated COX-2 protein and increased basal expression of cPLA$_2$ protein. We postulate that the significantly increased levels of COX-2 and cPLA$_2$ in COX-1$^{-/-}$ cells are likely to account for the increased rates of PGE$_2$ biosynthesis; these data also implicate the existence of compensatory mechanisms for PGE$_2$ production in COX-isoenzyme-deficient cells.

### *6.3.1.4 Different Substrate-Utilization Patterns in COX Knockout Cells*

To determine whether COX-1 and COX-2 have a preference for endogenous (Fig. 1A) or exogenous AA for conversion to PGE$_2$ and to verify that COX-1 was indeed expressed in COX-2$^{-/-}$ cells (as judged by its ability to synthesize PGE$_2$), we added exogenous [$^{14}$C]-AA to wild-type, COX-2$^{-/-}$ and COX-1$^{-/-}$ cells, and IL-1-induced PGE$_2$ production was compared in AA-supplemented cells and cells without added AA. The results shown in Fig. 4 show that both wild-type cells and cells expressing only COX-1 (COX-2$^{-/-}$) synthesized similar amounts of PGE$_2$ from exogenous AA, while cells only expressing COX-2 (COX-1$^{-/-}$) were essentially unable to synthesize PGE$_2$ from exogenous AA. Thus, agonists that induce PGE$_2$ biosynthesis in COX-2$^{-/-}$ cells in the absence of exogenous AA do so by mobilizing endogenous substrate (Fig. 1A). Based on these data, we conclude that, in COX-2$^{-/-}$ cells, substrate is likely to be the limiting reagent for constitutively expressed, COX-1-mediated PGE$_2$ biosynthesis. Figure 4 also shows that COX-1$^{-/-}$ cells are able to utilize both exogenous and endogenous substrates (Fig. 1A). However, IL-1, TNF and FGF significantly enhance the ability of COX-1$^{-/-}$ cells to produce PGE$_2$, most likely by enhancing COX-2 expression, as shown in Fig. 2. In addition, COX-1$^{-/-}$ cells treated with PMA did not produce elevated levels of PGE$_2$, even when exogenous AA was provided. This indicates that PMA probably increased PGE$_2$ production by increasing the availability of endogenous AA in COX-1$^{-/-}$ cells, while IL-1, TNF and FGF probably affect AA mobilization and COX-2 expression. PMA affected the wild-type cells similarly. These results clearly raise the possibility that, in COX-1 or -2 knockout cells, there is a co-ordinate upregulation of the expression and/or activities of COX-1, COX-2 and cPLA$_2$, leading to increased PGE$_2$ biosynthesis. These data also demonstrate that both COX-1$^{-/-}$ and COX-2$^{-/-}$

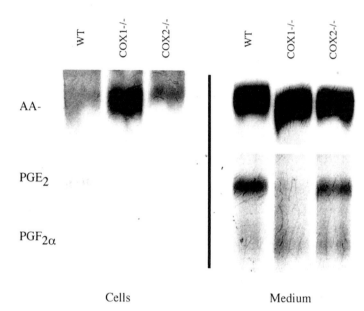

**Fig. 4.** Different substrate utilization patterns in cyclooxygenase (*COX*) knockout cells. Wild-type, COX-1$^{-/-}$ and COX-2$^{-/-}$ cells were stimulated with interleukin 1 in the presence of exogenous [$^{14}$C]-arachidonic acid. Labeled products from the cells and supernatant were analyzed separately using radio thin-layer chromatography

cells can effectively utilize AA from either endogenous or exogenous sources. Murakami et al. (1999) have recently reported that, indeed, there is a distinct functional coupling between various PLA$_2$s and the COX isoenzymes during both immediate and delayed prostanoid biosynthesis using COX-, PLA$_2$- and COX-isoenzyme-transfected cells.

### 6.3.1.5 Co-ordinate Regulation of COX-1, COX-2 and cPLA$_2$ Expression

Our data are consistent with the hypothesis that long-term COX-isoenzyme deficiency results in the altered expression of the remaining two enzymes that regulate mobilization and conversion of AA to PGs. Figure 5 summarizes the patterns of COX-1, COX-2 and cPLA$_2$ expression in response to IL-1 in COX-null cells compared with normal cells. The

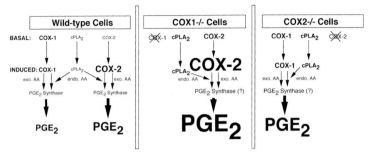

Fig. 5. The co-ordinate regulation of cyclooxygenase 1 (*COX-1*), COX-2 and cytosolic-phospholipase-$A_2$ expression in COX knockout cells. In wild-type cells, inducible COX-2 is responsible for the increase in cytokine-induced prostaglandin $E_2$ (*$PGE_2$*) production, as represented by proportionally larger characters and lines throughout the diagram. Although the constitutive level of COX-1 expression is low, exogenous AA (exo. AA) is effectively converted to $PGE_2$. In COX-$2^{-/-}$ or COX-$1^{-/-}$ cells, the overexpression of $cPLA_2$ may play a role in increased basal $PGE_2$ biosynthesis (compared with wild-type cells) by increasing the availability of endogenous AA. Cytokines greatly induce COX-2 accumulation in COX-$1^{-/-}$ cells, resulting in enhanced $PGE_2$ biosynthesis. In COX-$2^{-/-}$ cells, overexpression of COX-1 and $cPLA_2$ leads to an increase in basal $PGE_2$ biosynthesis (compared with wild-type cells). However, cytokines do not enhance $PGE_2$ biosynthesis in COX-$2^{-/-}$, due to the lack of increased COX-1 expression. As in wild-type cells, exo. AA is effectively utilized by COX-1 in COX-$2^{-/-}$ cells, as indicated by high levels of $PGE_2$ accumulation. At present, the effects of COX-isoenzyme deficiency on the expression of $PGE_2$ synthase are not known, but this enzyme does not appear to be rate limiting (Kirtikara et al. 1998)

scheme shows the compensatory expression of the alternative COX isoenzyme and $cPLA_2$ when one of the COX isoenzymes is absent. In cells lacking the housekeeping isoenzyme (COX-1), overcompensation results in the overexpression of COX-2 and $cPLA_2$ and, in turn, elevated $PGE_2$ biosynthesis. Similarly, cells lacking the inducible isoenzyme (COX-2) elicit enhanced expression of COX-1 and $cPLA_2$. While we are unable to comment on the precise status of $PGE_2$-synthase expression in wild-type, COX-$1^{-/-}$ or COX-$2^{-/-}$ cells, we have detected its expression in different cell types. Because $PGE_2$ is the predominant prostanoid product, its expression would not appear to be rate-limiting, given the great potential for $PGE_2$ biosynthesis in the presence of exogenous AA (Fig. 4). Thus, our data clearly show that COX-deficient cells have the potential to overcome the lack of expression of one or the other COX isoenzymes by overexpressing the alternate COX isoform and increasing $cPLA_2$ expression. Such a potential mechanism for production of $PGE_2$ by cells in vitro is not surprising, because neither COX-1- (Langenbach et al. 1995) nor COX-2-deficient mice (Morham et al. 1995) showed immediate postnatal mortality or severe developmental arrest in utero. However, in contrast to the results shown here using lung fibroblasts, Langenbach et al. (1995) did not report any compensatory COX-2-mediated $PGE_2$ production in glandular stomachs of COX-1-deficient mice, suggesting that tissue specificity may also be an important subject of further investigation. Together, these findings underscore the importance of elucidating the potential long-term effects of COX-1 or COX-2 inhibition with respect to alterations in the quantitative and/or qualitative patterns of AA metabolism.

### 6.3.1.6 Is There Compensatory COX Isoenzyme Expression In Vivo?

The short answer to this question is that we aren't sure – yet. At this point, the major clue that compensation is at least a possibility in vivo is that neither COX-1 nor COX-2 mice appear to show any gross physiological problems, and their life expectancy is the same as wild-type mice. In our studies, COX-$2^{-/-}$ mice do exhibit some problems with reproduction (reduced litter sizes) compared with wild-type and COX-1 knockout counterparts. However, our novel IA-q/COX-2 knockout mice, developed for their susceptibility to CIA by back-crossing to DBA1 mice, do not die from renal failure after 10–14 weeks and, on histological examination, our COX-$2^{-/-}$ mice appeared healthy and without gross kidney pathology. This observation is in stark contrast to the severe renal pathology described in the IAb-COX-$2^{-/-}$ mice originally developed by Morham et al. (1995) in C57/BL6 mice. This difference in kidney function in COX-$2^{-/-}$ mice with different genetic backgrounds suggests that both genetic background and COX deficiency play a role in kidney pathology. Other examples of compensation may include the possibility that our COX-1 knockout mice do not have bleeding problems because their platelets do not express COX-1. In order to address the issue of compensation at the molecular level, we are in the process of examining the relative expression levels of the COX isoforms in tissues from knockout mice.

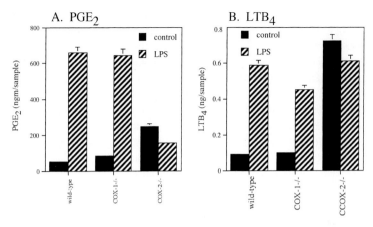

**Fig. 6.** The air-pouch-inflammation model in cyclooxygenase (*COX*) knockout mice. In this experiment, air pouches were made in wild-type, COX-1$^{-/-}$ and COX-2$^{-/-}$ mice, which were then challenged with either phosphate-buffer saline (PBS) or PBS/lipopolysaccharide injected into the pouch. After 6 h, the animals were sacrificed, and the infiltrates in each pouch were collected by lavage with PBS. Cells were removed and counted, and the exudate was analyzed for the presence of both prostaglandin $E_2$ and leukotriene $B_4$

### 6.3.1.7 Is There Compensatory Expression of Other Eicosanoid Generating Enzymes in Cox Knockout Mice?

While we have not directly examined the expression levels of (for example) 5-lipoxygenase in tissues from COX knockout mice, we have compared PG and LT synthesis using the mouse air-pouch model in an effort to examine compensation at the biochemical level. Our very preliminary results from these studies are shown in Fig. 6. In these experiments, air pouches were made in wild-type, COX-1$^{-/-}$ and COX-2$^{-/-}$ mice, which were then challenged with either phosphate-buffered saline (PBS) or PBS/LPS injected into the pouch. After 6 h, the animals were sacrificed, and the infiltrates in each pouch were collected by lavage. Cells were removed and counted, and the exudate was analyzed for the presence of both $PGE_2$ and $LTB_4$. As shown in Fig. 6A, $PGE_2$ production was quite similar in wild-type and COX-1$^{-/-}$ mice, both exhibiting significant increases in PG production in response to LPS. Conversely, in COX-2$^{-/-}$ mice, basal PG production was elevated compared with production in wild-type and COX-1$^{-/-}$ mice, and the response to LPS was totally ablated. These results indicate that COX-2 is they key enzyme regulating LPS-induced PG synthesis in this model. The data also suggest that COX-1 may be more highly constitutively expressed in COX-2 knockouts, accounting for the higher basal levels of PG in control animals. In Fig. 6B, LTB$_4$ levels in pouch exudates are determined. LTB$_4$ biosynthesis follows the same pattern as PG production in wild-type and COX-1$^{-/-}$ mice, with LPS inducing a strong increase in LTB$_4$ levels. However, in COX-2$^{-/-}$ mice, basal LTB$_4$ production was dramatically elevated in control animals, and there was no increase in LPS-stimulated LTB$_4$ production. These results suggest that, when COX-2 is knocked out, COX-1 expression may be increased; lipoxygenase activity may also be increased, suggesting that there may be cross-talk between the COX- and LT-biosynthesis pathways. Of course, these results could also be explained by the fact that there may be more substrate available for LT biosynthesis in cells that do not express COX-2. Again, these are very preliminary data, and studies are currently underway to determine the mechanisms regulating enhanced LT production in COX-2$^{-/-}$ mice and in in vitro studies using cells (polymorphonuclear neutrophils) that express lipoxygenase. These studies are analogous to those using COX-knockout-mouse lung cells, as described above.

### 6.3.2 COX and Fever in Knockout Mice

#### 6.3.2.1 Effects of LPS on the Core Temperature of Wild-Type Mice

The i.p.-injection procedure and associated handling rapidly induce a transient rise of approximately 1°C in the T$_c$s of wild-type mice despite their training. However, over the following 45 min, it abated in the PFS-treated group (Fig. 7, *open circles*). However, LPS administration prevented this recovery, as evidenced by the sustained 1°C T$_c$ rise over the first 1.5 h (Fig. 7, *closed circles*). The fever gradually abated over the next 2.5 h. The responses to PFS and LPS treatments were significantly different ($P<0.0001$, $F=352.29$).

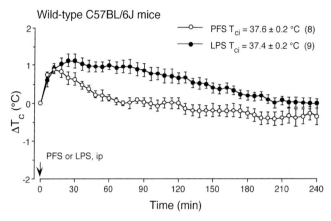

**Fig. 7.** Effects of intraperitoneally injected pyrogen-free saline (*PFS*, 0.2 ml; *open circles*) and *Escherichia coli* lipopolysaccharide (*LPS*; 1 µg/mouse in 0.2 ml PFS; *closed circles*) on the core temperature ($T_c$) of wild-type C57BL/6J mice. A 3-h stabilization period preceded the collection of these data. The $T_c$s are expressed as differences ($\Delta T_c$) relative to their initial levels ($T_{ci}$, the average of the $T_c$s over the last 10 min before the injection of PFS or LPS). The values are means ± standard errors; *numbers in parentheses* are the numbers of animals (Li et al. 1999)

### 6.3.2.2 Effects of LPS on the Core Temperature of Wild-Type, COX-1$^{+/-}$ and COX-1$^{-/-}$ Mice

No difference in basal $T_c$ was observed between the wild-type and COX-1 transgenic mice. The i.p. injection of PFS induced no demonstrable thermal effect (other than that associated with handling and the injection itself) in any group (Fig. 8, top). In both the COX-1$^{+/-}$ and COX-1$^{-/-}$ mice, LPS administration induced febrile responses that were not different than those evoked in the wild-type mice (Fig. 8, bottom). The maximum $T_c$ elevation was on the order of 1°C and was reached early, then decreased very slowly over the 4-h duration of the measurements.

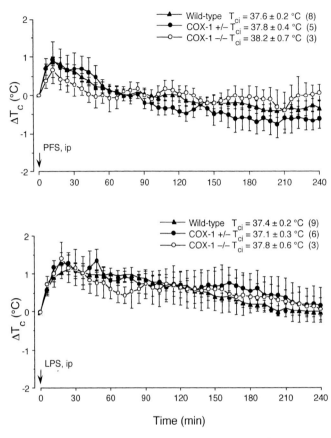

**Fig. 8.** Effects of intraperitoneally injected pyrogen-free saline (*PFS*, 0.2 ml; top) and *Escherichia coli* lipopolysaccharide (*LPS*; 1 µg/mouse in 0.2 ml PFS; bottom) on the core temperature ($T_c$) of wild-type (*triangles*), cyclooxygenase 1 (COX-1)$^{+/-}$ (*closed circles*) and COX-1$^{-/-}$ (*open circles*) C57BL/6J mice. A 3-h stabilization period preceded the collection of these data. The $T_c$s are expressed as differences ($\Delta T_c$) relative to their initial levels ($T_{ci}$, the average of the $T_c$s over the last 10 min before the injection of PFS or LPS). The values are means ± standard errors; *numbers in parentheses* are the numbers of animals (Li et al. 1999)

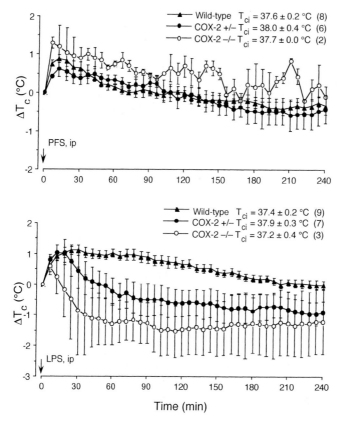

**Fig. 9.** Effects of intraperitoneally injected pyrogen-free saline (*PFS*, 0.2 ml; top) and *Esherichia coli* lipopolysaccharide (*LPS*; 1 µg/mouse in 0.2 ml PFS; bottom) on the core temperature ($T_c$) of wild-type (*triangle*), cyclooxygenase 2 (COX-2)$^{+/-}$ (*closed circles*), and COX-2$^{-/-}$ (*open circles*) C57BL/6J mice. A 3-h stabilization period preceded the collection of these data. The $T_c$s are expressed as differences ($\Delta T_c$) relative to their initial level ($T_{ci}$, the average of the $T_c$s over the last 10 min before the injection of PFS or LPS). The values are means ± standard errors; *numbers in parentheses* are the numbers of animals (Li et al. 1999)

### 6.3.2.3 Effects of LPS on the Core Temperature of Wild-Type, COX-2$^{+/-}$ and COX-2$^{-/-}$ Mice

No difference in basal $T_c$ was observed between the wild-type and COX-2 transgenic mice, and PFS i.p. induced no demonstrable thermal effect (other than that associated with handling and the injection) in any group (Fig. 9, top). The $T_c$s of the COX-2$^{-/-}$ mice, however, tended to be more labile than those of the other mice. LPS administration caused approximately 0.5°C and 1.4°C falls below the basal values of the $T_c$s of the COX-2$^{+/-}$ and COX-2$^{-/-}$ mice, respectively, in contrast to its 1°C fever-producing effect in the wild-type mice. However, the overall fever curves of the two groups were not statistically significantly different from each other. The lowest levels in both transgenic groups were attained in approximately 60 min and persisted without recovery for the remainder of the experimental period (Fig. 9, bottom). The responses of the wild-type and COX-2$^{+/-}$ and COX-2$^{-/-}$ mice were significantly different ($P<0.0001$, $F=248.93$ and $331.80$, respectively). The $T_c$ falls in the transgenic mice occurred after the initial $T_c$ rises provoked by the handling procedures.

### 6.3.2.4 The Fever Indices Induced by LPS in Wild-Type, COX-1 and COX-2 Knockout Mice

The fever indices of the wild-type, COX-1$^{+/-}$ and COX-1$^{-/-}$ mice were not significantly different. However, the fever indices of the COX-2$^{+/-}$ and COX-2$^{-/-}$ mice were significantly lower than those of the wild-type mice, though they were not different from each other (Fig. 10, *closed bars*). Nevertheless, taken together, the COX-2 extant in the brains of the transgenic mice appeared to influence the severity of their fevers, as evaluated by the correlation between the fever indices induced by LPS and the COX-2 rating values. These rating values are estimated on a scale of zero to three, based on the density of immunostaining of COX-2 in cerebrovascular venular endothelial cells (see below) of wild-type (index=3) and COX-2-gene knockout (index<3) mice.

### 6.3.2.5 COX Immunostaining in Brain Tissue

Immunohistochemical staining confirmed that the mice used in the present study were properly grouped according to their genotypes (Fig. 11). COX-1-like immunoreactivity was found in glia-like small cells and some neurons in the brains of wild-type and COX-2$^{-/-}$ mice

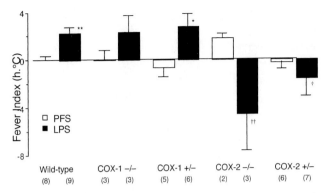

**Fig. 10.** The fever indices induced by lipopolysaccharide (*LPS*; 1 μg/mouse, intraperitoneally) in wild-type and cyclooxygenase 1 and 2 gene knockout C57BL/6J mice. The values are means ± standard errors; *numbers in parentheses* are the numbers of animals. *Single asterisks* indicate P<0.05, and *doubled asterisks* indicate P<0.01 (compared with the corresponding pyrogen-free-saline-treated group or the wild-type group that received LPS; Li et al. 1999)

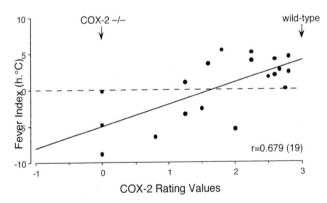

**Fig. 11.** Correlation between the fever indices induced by lipopolysaccharide- and cyclooxygenase 2 (*COX-2*). Rating values are estimated on a scale of zero to three, based on the density of immunostaining of COX-2 in cerebrovascular venular endothelial cells of wild-type (index=3) and COX-transgenic (index<3) mice (Li et al. 1999)

but not in those of COX-1 transgenics. The staining of the glia-like cells was more intense than that of the neurons. LPS treatment did not change the number of COX-1-like immunoreactive cells or the intensity of their staining. Constitutive COX-2-like immunoreactivity was found in the cortical neurons of wild-type and COX-1$^{-/-}$ mice but not in those of COX-2$^{-/-}$ mice. LPS treatment strongly induced COX-2-like immunoreactivity in the endothelial cells of the parenchymal and subarachnoidal blood vessels. COX-2-like immunoreactivity in neurons was not enhanced. No such staining was present in LPS-treated COX-2$^{-/-}$ mice.

Taken together, these results demonstrate that COX-2-gene-deficient mice are unable to develop a full fever in response to the i.p. administration of a pyrogenic dose of LPS. The extent of the diminution of the febrile response appears to be proportional to the reduction in COX-2 expressed by these animals, as visualized by immunostaining of their cerebrovascular venular-endothelium cells. The more moderate LPS-induced fall in the $T_c$s seen in COX-2$^{+/-}$ mice compared with that in COX-2$^{-/-}$ mice may reflect a gene-dosing effect, but this is speculative. These data also show that, in contrast, wild-type, COX-1$^{+/-}$ and COX-1$^{-/-}$ mice all respond to LPS with an approximately 1°C rise in $T_c$ within 1 h after i.p. injection of LPS. Therefore, it can be concluded from these observations that COX-2 is critically important for LPS-induced fever production in mice.

This is not an unexpected finding, however. Indeed, a substantial body of evidence implicating COX-2 specifically as an essential factor in fever production has accumulated over the last several years. For example, it was shown that COX-2-selective inhibitors suppress the $T_c$ rise produced by i.p. and intracerebroventricular injections of LPS, IL-1β and TNFα (Cao et al. 1996, 1997, 1998) and that LPS and TNFα (by any route of administration) induce COX-2 messenger RNA and protein in brain tissue (Elmquist et al. 1997a). It was also shown that the induction of COX-2 protein precedes the onset of fever after i.p. (but not i.v.) injection of LPS in rats and that the duration of fever correlates with the number of COX-2-positive endothelial cells (Matsumura et al. 1998a). This congruence between COX-2 expression and fever suggests that the pyrogen-induced elevation of $PGE_2$ may be accounted for by COX-2 under these experimental conditions. Taken together, these data seem to provide compelling support for the pivotal role of COX-2 in fever genesis.

It remains controversial, however, whether the affected COX-2-positive cells are endothelial cells (Van Dam et al. 1993; Cao et al. 1996, 1998; Matsumura et al. 1998; Quan et al. 1998), neurons (Van Dam et al. 1993; Breder 1997), perivascular microglia or meningeal macrophages (Elmquist et al. 1997a). The various viewpoints have recently been reviewed by Matsumura et al. (1998b) and by Elmquist et al. (1997b). The present results support the brain vasculature as the site of synthesis of $PGE_2$ associated with fever. Although COX-2 was also expressed in non-stimulated neurons of wild-type and COX-1 transgenic mice, it was evidently not upregulated by the present pyrogenic stimulus. COX-2-like immunoreactivity was also not evident in perivascular cells, though the time after the injection of LPS when these brains were examined (~4 h) may have been a factor (Quan et al. 1998).

The febrile courses observed in the present study were similar to those reported by Wang et al. (1997) under nearly analogous experimental conditions. They differed, however, from those obtained by others, e.g., Kozak et al. (1994). It is likely that differences in LPS doses, ambient temperature, mouse strain and methods of $T_c$ recording accounted for the apparent discrepancy. It also appeared that the COX-2$^{-/-}$ mice were warmer than the wild-types following PFS administration. However, the number of homozygotes available for this comparison was too small to allow us to be certain of this effect. Finally, it should be noted that the $T_c$s of both the COX-2-gene-deficient groups fell below their initial levels following LPS administration instead of simply returning to those levels. It is intriguing that such a reversal seems to occur whenever an apparently key element in the mediation of the febrile response to LPS is eliminated (Sehic et al. 1996). This is entirely speculative, but it may be that, in the absence of a function essential to driving the febrile response, other factors also activated by LPS that might normally mitigate the febrile rise are freed to exert their counter-regulatory, antipyretic effect.

It may be of additional interest that, while neither the COX-2$^{+/-}$ nor the COX-2$^{-/-}$ mice produced febrile responses to LPS, during handling they nevertheless exhibited prototypic initial, transient increases in $T_c$ that were not different from those of their wild-type counterparts or their COX-1-deficient analogs. It has been suggested that, in rats, the rise in

$T_c$ resulting from handling is a fever-like, $PGE_2$-mediated stress or anxiogenic response, because NSAIDs (salicylate) prevent it (Briese and Cabanac 1991). However, in mice, this effect may not be a fever, because the magnitude of the handling-associated $T_c$ rise is reportedly not diminished by salicylate administration; however, the $T_ci$ of the treated animals is lower (Cabanac and Briese 1991). The present data seem to corroborate the previous findings in mice, but only in part; the transient hyperthermia associated with handling was not demonstrably affected by the deletion of either the COX-1 or COX-2 gene in these mice. However, analogous to the reported effect of COX-2-selective inhibitors (Futaki et al. 1994; Chan et al. 1995; Cao et al. 1998), the basal $T_c$s of these mice were not reduced. It would seem, therefore, that the stress hyperthermia provoked in mice by the handling procedures may not involve PGs.

Since, like wild-type mice, COX-$1^{+/-}$ and COX-$1^{-/-}$ mice responded to i.p. LPS with fevers whereas COX-$2^{+/-}$ and COX-$2^{-/-}$ mice displayed no fever after LPS challenge, it would appear that COX-2 is critically important for LPS-induced fever production in these animals. However, COX-1- and COX-2-gene deletions do not seem to affect the basal $T_c$s or handling-stress hyperthermia.

## 6.4 Concluding Remarks

Evidence from COX knockout cells suggests that the ablation of expression of one of the COX isoenzymes results in the overexpression of the alternative COX isoform. Evidence also suggests that other cellular components involved in eicosanoid biosynthesis, such as $PLA_2$ (substrate mobilization) and 5-lipoxygenase may be modulated by the absence of one of the COX isoenzymes. Studies designed to determine whether "compensation" or cross-talk between eicosanoid-biosynthesis pathways occurs in response to the lack of one or the other COX isoenzymes in vivo are underway. As for thermal regulation, it appears that COX-2 is critically important for LPS-induced fever production. However, neither COX-1- nor COX-2-gene deletion seem to affect basal $T_c$s or handling-stress hyperthermia. In this instance, it is clear that one COX isoform cannot functionally substitute for the other COX isoform, regardless of the level of expression. This suggests that both the level of expression and the tissue and/or functional specificity must be considered with respect to COX-isoform expression and isoform roles in various physiological processes.

## References

Blatteis C, Schic E (1997) Prostaglandin $E_2$: a putative fever mediator. In: Mackowiak PA (ed) Fever: basic mechanisms and management, 2nd edn. Lippincott-Raven, New York, pp 260–266

Boulant JA (1996) Hypothalamic neurons regulating body temperature. In: Fregley MJ, Blatteis CM (eds) Handbook of physiology. Oxford, New York, pp 105–126

Breder CD (1997) Cyclooxygenase systems in the mammalian brain. Ann N Y Acad Sci 813:296–301

Breder CD, Smith WL, Raz A, Masferrer J, Seibert K, Needleman P, Saper CB (1992) Distribution and characterization of cyclooxygenase immunoreactivity in the ovine brain. J Comp Neurol 322:409–438

Breder CD, Dewitt D, Kraig RP (1995) Characterization of inducible cyclooxygenase in rat brain. J Comp Neurol 355:296–315

Briese E, Cabanac M (1991) Stress hyperthermia: physiological arguments that it is a fever. Physiol Behav 49:1153–1157

Cabanac A, Briese E (1991) Handling elevates the colonic temperature of mice. Physiol Behav 199:95–98

Cao C, Matsumura K, Yamagata K, Watanabe Y (1996) Endothelial cells of the rat brain vasculature express cyclooxygenase-2 mRNA in response to systemic interleukin-1β: a possible site of prostaglandin synthesis responsible for fever. Brain Res 733:263–272

Cao C, Matsumura K, Yamagata K, Watanabe Y (1997) Involvement of cyclooxygenase-2 in LPS-induced fever and regulation of its mRNA by LPS in he rat brain. Am J Physiol 272:R1722–R1725

Cao C, Matsumura K, Yamagata K, Watanabe Y (1998) Cyclooxygenase-2 is induced in brain blood vessels during fever evoked by peripheral or central administration of tumor necrosis factor. Brain Res Mol Brain Res 56:45–56

Chan CC, Boyce S, Brideau C, Ford-Hutchinson AW, Gordon R, Guay D, Hill RG, Li CS, Mancini J, Penneton M, et al. (1995) Pharmacology of a selective cyclooxygenase-2 inhibitor, L-745,337: a novel nonsteroidal anti-inflammatory agent with an ulcerogenic sparing effect in rat and nonhuman primate stomach. J Pharmacol Exp Ther 274:1531–1537

Clark JD, Lin LL, Kriz RW, Ramesha CS, Sultzman LA, Lin AY, Milona N, Knopf JL (1991) A novel arachidonic acid-selective cytosolic $PLA_2$ contains a $Ca^{(2+)}$-dependent translocation domain with homology to PKC and GAP. Cell 65:1043–1051

Coceani F (1991) Prostaglandins and fever: facts and controversies. In: Mackowiak PA (ed) Fever: basic mechanisms and management, 2nd edn. Lippincott-Raven, New York, pp 59–70

Coceani F, Bishai I, Dinarello CA, Fitzpatrick FA (1983) Prostaglandin $E_2$ and thromboxane $B_2$ in cerebrospinal fluid of afebrile and febrile cat. Am J Physiol 244:R785–R793

Coceani F, Lees J, Bishai I (1988) Further evidence implicating prostaglandin $E_2$ in the genesis of pyrogen fever. Am J Physiol 254:R463–R469

DeWitt DL, Meade EA, Smith WL (1993) PGH synthase isozyme selectivity: potential safer nonsteroidal anti-inflammatory drugs. Am J Med 95:40S–44S

Eguchi N, Hayashi H, Urade Y, Ito S, Hayaishi O (1988) Central action of prostaglandin $E_2$ and its methylester in the induction of hyperthermia after their systemic administration in urethane-anesthetized rats. J Pharmacol Exp Ther 247:671–679

Elmquist JK, Breder CD, Sherin JE, Scammell TE, Hickey WF, Dewitt D, Saper CB (1997a) Intravenous lipopolysaccharide induces cyclooxygenase-2-like immunoreactivity in rat brain perivascular microglia and meningeal macrophages. J Comp Neurol 381:119–129

Elmquist JK, Scammell TE, Saper CB (1997b) Mechanisms of CNS response to systemic immune challenge: the febrile response. Trends Neurosci 20:565–570

Futaki N, Takahashi S, Yokoyama M, Arai I, Higuchi S, Otomo S (1994) NS-398, a new anti-inflammatory agent, selectively inhibits prostaglandin G/H synthase/cyclooxygenase (COX-2) activity in vitro. Prostaglandins 47:55–59

Goetzl EJ, An S, Smith WL (1995) Specificity of expression and effects of eicosanoid mediators in normal physiology and human diseases. FASEB J 9:1051–1058

Herschman H (1996) Prostaglandin synthase 2. Biochim Biophys Acta 1299:125–140

Herschman HR, Gilbert RS, Xie W, Luner S, Reddy ST (1995) The regulation and role of TIS10 prostaglandin synthase-2. Adv Prostaglandin Thromboxane Leukot Res 23:23–28

Kargman SL, O'Neill GP, Vickers PJ, Evans JF, Mancini JA, Jothy S (1995) Expression of prostaglandin G/H synthase-1 and -2 protein in human colon cancer. Cancer Res 55:2556–2559

Kaufmann WE, Worley PF, Pegg J, Bremer M, Isakson P (1996) COX-2, a synaptically induced enzyme, is expressed by excitatory neurons at postsynaptic sites in rat cerebral cortex. Proc Natl Acad Sci U S A 93:2317–2321

Kirtikara K, Morham SG, Raghow R, Laulederkind SJ, Kanekura T, Goorha S, Ballou LR (1998) Compensatory prostaglandin $E_2$ biosynthesis in cyclooxygenase 1 or 2 null cells. J Exp Med 187:517–523

Kozak W, Conn CA, Kluger MJ (1994) Lipopolysaccharide induces fever and depresses locomotor activity in unrestrained mice. Am J Physiol 266:R125–R135

Langenbach R, Morham SG, Tiano HF, Loftin CD, Ghanayem BI, Chulada PC, Mahler JF, Lee CA, Goulding EH, Kluckman KD, et al. (1995) Prostaglandin synthase 1 gene disruption in mice reduces arachidonic acid-induced inflammation and indomethacin-induced gastric ulceration. Cell 83:483–492

Li S, Wang Y, Matsumura K, Ballou LR, Morham SG, Blatteis CM (1999) The febrile response to lipopolysaccharide is blocked in cyclooxygenase-$2^{-/-}$, but not in cyclooxygenase-$1^{-/-}$ mice. Brain Res 825:86–94

Lin LL, Lin AY, Knopf JL (1992) Cytosolic phospholipase $A_2$ is coupled to hormonally regulated release of arachidonic acid. Proc Natl Acad Sci U S A 89:6147–6151

Matsumura K, Cao C, Ozaki M, Morii H, Nakadate K, Watanabe Y (1998a) Brain endothelial cells express cyclooxygenase-2 during lipopolysaccharide-induced fever: light and electron microscopic immunocytochemical studies. J Neurosci 18:6279–6287

Matsumura K, Cao C, Watanabe Y, Watanabe Y (1998b) Prostaglandin system in the brain: sites of biosynthesis and sites of action under normal and hyperthermic states. Prog Brain Res 115:275–295

Milton AS (1990) Is the prostaglandin $E_2$ responsible for pyrogen fever centrally or peripherally derived? Acta Physiol Pol 41:9–17

Morham SG, Langenbach R, Loftin CD, Tiano HF, Vouloumanos N, Jennette JC, Mahler JF, Kluckman KD, Ledford A, Lee CA, et al. (1995) Prostaglandin synthase 2 gene disruption causes severe renal pathology in the mouse. Cell 83:473–482

Morimoto A, Morimoto K, Watanabe T, Sakata Y, Murakami N (1992) Does an increase in prostaglandin $E_2$ in the blood circulation contribute to a febrile response in rabbits? Brain Res Bull 29:189–192

Murakami M, Kambe T, Shimbara S, Kudo I (1999) Functional coupling between various phospholipase $A_2$s and cyclooxygenases in immediate and delayed prostanoid biosynthetic pathways. J Biol Chem 274:3103–3115

O'Neill GP, Ford-Hutchinson AW (1993) Expression of mRNA for cyclooxygenase-1 and cyclooxygenase-2 in human tissues. FEBS Lett 330:156–160

Quan N, Whiteside M, Herkenham M (1998) Cyclooxygenase-2 mRNA expression in rat brain after peripheral injection of lipopolysaccharide. Brain Res 802:189–197

Reddy ST, Herschman HR (1996) Transcellular prostaglandin production following mast cell activation is mediated by proximal secretory phospholipase $A_2$ and distal prostaglandin synthase 1. J Biol Chem 271:186–191

Rigas B, Goldman IS, Levine L (1993) Altered eicosanoid levels in human colon cancer. J Lab Clin Med 122:518–523

Roberts LJ, Morrow JD (1997) The generation and actions of isoprostanes. Biochim Biophys Acta 1345:121–135

Rosloneic E (1999) Molecular biology of autoimmune arthritis. In: Serhan C, Weller P (eds) Molecular and cellular basis of inflammation. Humana, Totowa, pp 19–99

Rotondo D, Abul HT, Milton AS, Davidson J (1988) Pyrogenic immunomodulators increase the level of prostaglandin $E_2$ in the blood simultaneously with the onset of fever. Eur J Pharmacol 154:145–152

Sehic E, Szekely M, Ungar AL, Oladehin A, Blatteis CM (1996) Hypothalamic prostaglandin $E_2$ during lipopolysaccharide-induced fever in guinea pigs. Brain Res Bull 39:391–399

Smith WL, Garavito RM, DeWitt DL (1996) Prostaglandin endoperoxide H synthases (cyclooxygenases)-1 and -2. J Biol Chem 271:33157–33160

Ushikubi F, Segi E, Sugimoto Y, Murata T, Matsuoka T, Kobayashi T, Hizaki H, Tuboi K, Katsuyama M, Ichikawa A, Tanaka T, Yoshida N, Narumiya S (1998) Impaired febrile response in mice lacking the prostaglandin E receptor subtype $EP_3$. Nature 395:281–284

Van Dam A-M, Brouns M, Man-A-Hing W, Berkenbosch F (1993) Immunocytochemical detection of prostaglandin $E_2$ in microvasculature and in neurons of rat brain after administration of bacterial endotoxin. Brain Res 613:331–336

Vane JR, Botting RM (1994) Regulatory mechanisms of the vascular endothelium: an update. Pol J Pharmacol 46:499–521

Vane JR, Botting RM (1995a) New insights into the mode of action of anti-inflammatory drugs. Inflamm Res 44:1–10

Vane JR, Botting RM (1995b) A better understanding of anti-inflammatory drugs based on isoforms of cyclooxygenase (COX-1 and COX-2). Adv Prostaglandin Thromboxane Leukot Res 23:41–48

Wang J, Ando T, Dunn AJ (1997) Effect of homologous interleukin-1, interleukin-6 and tumor necrosis factor-$\alpha$ on the core body temperature of mice. Neuroimmunomodulation 4:230–236

Yamagata K, Andreasson KI, Kaufmann WE, Barnes CA, Worley PF (1993) Expression of a mitogen-inducible cyclooxygenase in brain neurons: regulation by synaptic activity and glucocorticoids. Neuron 11:371–386

# 7 Leukotriene-B$_4$ Receptor and Signal Transduction

T. Shimizu, T. Yokomizo, and T. Izumi

| | | |
|---|---|---|
| 7.1 | Biosynthesis and Metabolism of Leukotrienes | 125 |
| 7.2 | Isolation of LTB$_4$ Receptor | 128 |
| 7.3 | Characterization of BLT | 131 |
| 7.4 | Primary Structures of BLT from Various Animals | 132 |
| 7.5 | Chemotactic Activity of LTB4 and BLT | 134 |
| 7.A | Appendix: Cys-LT1 Receptor and Signal Transduction | 136 |
| References | | 137 |

## 7.1 Biosynthesis and Metabolism of Leukotrienes

When cells are stimulated with various agonists, cells release arachidonic acid by the action of phospholipases, and arachidonic acid is further converted to various types of leukotrienes (LTs; Samuelsson et al. 1987; Shimizu and Wolfe 1990; Serhan et al. 1996). Recent findings from our laboratories and others have shown that cytosolic phospholipase A$_2$ (cPLA$_2$) plays a dominant role in arachidonate release in various cells (Bonventre et al. 1997; Leslie 1997; Uozumi et al. 1997). We isolated arachidonate 5-lipoxygenase from potato tubers for the first time and found that a single enzyme catalyzes both lipoxygenation and dehydration to yield LTA$_4$ (Shimizu et al. 1984). The dual enzyme activity of the 5-lipoxygenase was later confirmed by purification and complementary DNA (cDNA) cloning of mammalian 5-lipoxygenase (Rouzer et al. 1985, 1986; Shimizu et al. 1986; Ueda et al. 1986; Dixon

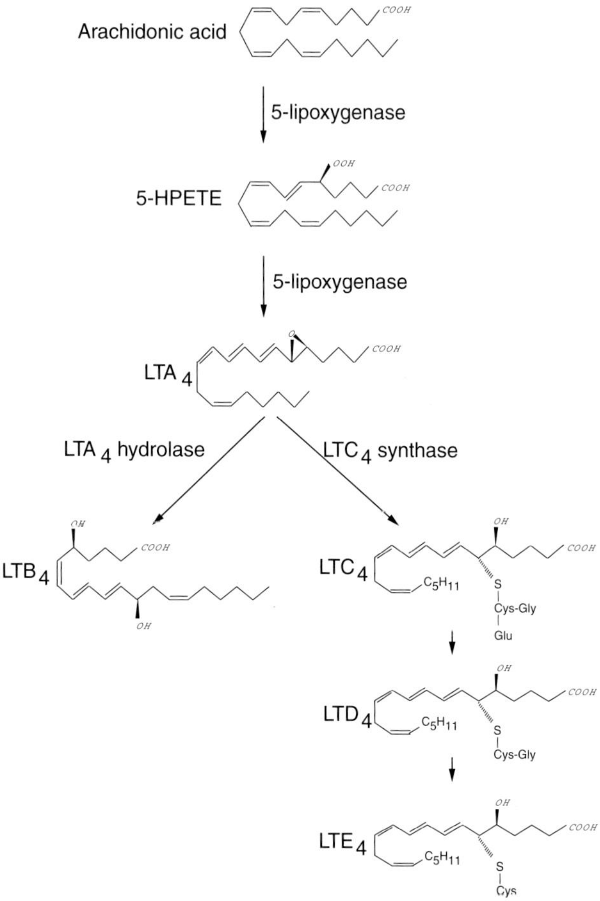

**Fig. 1.** Biosynthesis of leukotriene $B_4$ (*$LTB_4$*). Arachidonic acid is converted to $LTA_4$ by 5-lipoxygenase. $LTA_4$ is further converted to $LTB_4$ by $LTA_4$ hydrolase and to $LTC_4$ by $LTC_4$ synthase. $LTC_4$, $LTD_4$ and $LTE_4$ are collectively termed cysteinyl-LTs. *5-HPETE*, 5-hydroperoxy-eicosatetraenoic acid

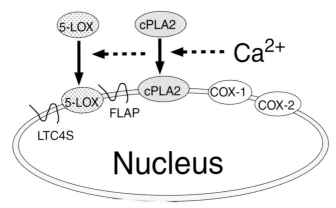

**Fig. 2.** Localization of enzymes involved in eicosanoid biosynthesis. Cytosolic phospholipase $A_2$ (*cPLA2*) and 5-lipoxygenase (*5-LOX*) cause translocation to the perinuclear region by increase in the intracellular $Ca^{2+}$ concentration (for cPLA2; Hirabayashi et al. 1999). COX (cyclo-oxygenase), leukotriene-$C_4$ synthase and 5-LOX-activating protein are constitutively present in the nuclear membrane and endoplasmic reticulum

et al. 1988; Chen et al. 1993). $LTA_4$ is further converted to $LTB_4$ by a cytosolic $LTA_4$ hydrolase (Rådmark et al. 1984; Minami et al. 1987; Ohishi et al. 1990) or to $LTC_4$ by the action of microsomal $LTC_4$ synthase (Fig. 1; Lam et al. 1994; Welsch et al. 1994). Recent findings suggest that all these enzymes are localized at the perinuclear membrane rather than the plasma membrane (Fig. 2), suggesting a novel role for these eicosanoids in intracellular or intranuclear actions. In the last decade, we have been extensively investigating how $LTB_4$ is biosynthesized and how it is inactivated. During the search for $LTB_4$ metabolism, we found that while $LTB_4$ is metabolized to ω-oxidized $LTB_4$ in polymorphonuclear leukocytes (Soberman et al. 1985; Kikuta et al. 1993), it is transformed into other products (Kumlin and Dahlén 1990) in most other tissues. Two metabolites (12-keto-$LTB_4$ and 12-keto-11, 12,14,15-tetrahydro-$LTB_4$) were identified by mass spectrometry and nuclear magnetic resonance and were found to be 1% as biologically active as $LTB_4$ (Yokomizo et al. 1993, 1996). Ensor et al. (1998) demonstrated that $LTB_4$ dehydrogenase is identical to 15-keto-prostaglandin 13-reductase. The two metabolic pathways of $LTB_4$ are illustrated in Fig. 3.

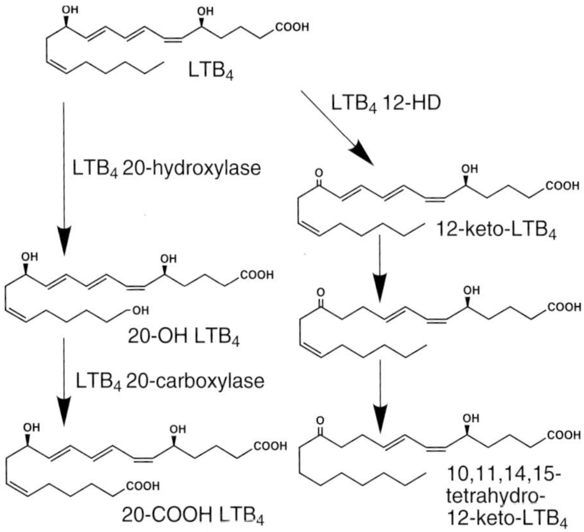

**Fig. 3.** Two alternative pathways of metabolic inactivation of leukotriene $B_4$ ($LTB_4$). In leukocytes, $LTB_4$ is metabolized to 20-OH-$LTB_4$ and 20-COOH-$LTB_4$. The pathways are not seen in other tissues. In the liver and kidney, $LTB_4$ is transformed into a 12-keto compound. An intermediate between 12-keto-$LTB_4$ and 10,11,14,15-tetrahydro-12-keto-$LTB_4$ has not been identified

## 7.2 Isolation of LTB$_4$ Receptor

LTs are reported to exert their biological activities thorough a membrane-spanning G-protein-coupled receptor (GPCR), though one report showed a nuclear $LTB_4$ receptor (Devchaund et al. 1996). These GPCRs are currently classified into three classes: $LTB_4$ receptor (BLT), cysteinyl (Cys)-LT1 and Cys-LT2. We recently isolated (using a subtraction strategy) a GPCR we later identified as BLT using retinoic acid-differentiated HL-60 cell cDNA libraries (Yokomizo et al. 1997). The receptor was originally misidentified as an atypical purinergic receptor (Akbar et al. 1996) and an orphan receptor expressed in leukocytes (Owman et al. 1996). Subtraction technology is illustrated in Fig. 4. We identified over 100 gene fragments using this method, including transcriptional factors, translational factors, cytochrome P450s, G-protein(s) and a GPCR (Table 1). Even using an expressed sequence tag database, 40

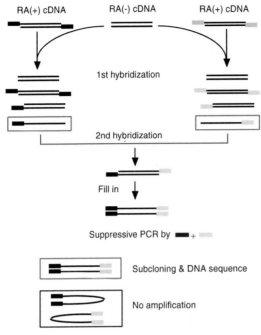

**Fig. 4.** Subtraction technology. Complementary DNA (cDNA) from retinoic acid (*RA*)-differentiated HL-60 cells was subtracted from that of undifferentiated cells. Each cDNA was digested by a four-base cutter (*Rsa I*) and was blunt-ended prior to subtraction. Single-stranded tester cDNAs ($RA^+$) in separate tubes were ligated to specially designed linkers (*black and gray boxes*), and hybridized to single-stranded driver cDNAs ($RA^-$). The second hybridization was performed in one tube without prior heat denaturing. After blunting by a Klenow fragment, RA-specific cDNAs were amplified by a suppressive polymerase chain reaction (*PCR*). In this case, only the cDNA fragments with different linkers on the 5' and 3' ends can be amplified. The DNA fragments with the same linkers on both sides cannot be amplified by PCR, because these linkers are self-complementary. The amplified DNA was sequenced, and homologous genes in the database were searched for by the Blast program. The method is slightly modified from the original protocol in the user's manual of Clonetech, Inc.

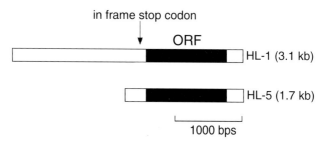

**Fig. 5.** Two leukotriene-B₄ receptor clones with identical open reading frame (*ORFs*). Two clones were isolated from a differentiated HL-60 cell library. HL-1 and HL-5 have identical ORFs

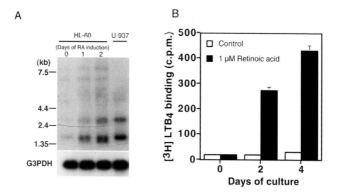

**Fig. 6.** Increase of leukotriene-B₄ (*LTB₄*) receptor levels in HL-60 cells. **A** Northern blotting showed two intense bands and a band with a size of approximately 7 kb. U-937 human histiocytic leukemia cells constitutively express the gene. **B** [³H]LTB₄ binding in HL-60 cells

genes are still unidentified. Using the fragment of the receptor as a probe, we screened differentiated HL-60 libraries and found two clones encoding GPCRs with an exactly identical open reading frame (Fig. 5). This must be a splicing variant, although we have to await the complete genomic analysis. We then performed a Northern blot and found that the receptor is highly expressed in human leukocytes, and some faint bands

**Table 1.** Retinoic acid-induced genes in HL-60 cells

G-protein-coupled receptor (BLT)
Transcriptional factors (TF-IIE, CDEI-binding protein)
Translational factor (elongation factor $1\alpha$)
Cytochrome P450s (CYP1B1, CYP2C11)
G-proteins ($\alpha\beta$ subunits)
Properdin, epithelial cell-growth factor
Unknown genes (>50)

*BLT*, leukotriene-B$_4$ receptor; *CDEI*, centromere DNA-sequence element I.

were also seen in the thymus and spleen. The increase in the intensity of two messenger RNA bands in HL-60 cells, together with the increase in LTB$_4$ binding during retinoic acid-induced differentiation (Fig. 6), suggested that the GPCR encodes BLT.

## 7.3 Characterization of BLT

We then transiently transfected the gene into Cos-7 cells and HEK-293 cells and found that the membranes from the transfected cells exhibited a specific and saturable binding to LTB$_4$. A $K_d$ value of 0.15 nM was obtained by Scatchard analyses; this is comparable with that seen in HL-60 cells and previously reported values (Miki et al. 1990; Kasimir et al. 1991). Several lines of Chinese-hamster ovary (CHO) cells stably expressing the GPCR gene showed that LTB$_4$ induces a calcium (Ca) increase and inhibits adenylate cyclase in pertussis-toxin (PTX)-insensitive and -sensitive manners, respectively. Since the cloning of BLT is a long sought goal for LT researchers and the gene was originally reported as a purinergic receptor (P2Y7), we then performed careful Ca-monitoring assays using a subline of C6 glioma cells (C6–15) that lack the intrinsic purinergic receptors. We transfected the BLT gene into the cells and (as shown in Fig. 7) adenosine triphosphate and its analogs did not show any Ca response, whereas LTB$_4$ induced Ca signals. All these results indicate that the cloned GPCR really encodes BLT.

The identity of the gene as BLT was confirmed by other groups (Owman et al. 1997; Huang et al. 1998), and a correction paper came out from the nomenclature committee of purinergic receptors (Khakh and

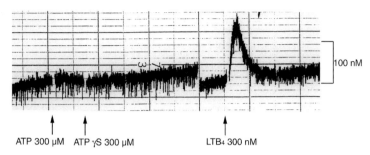

**Fig. 7.** Calcium response in C6–15 glioma cells. A subline of C6–15 glioma cells lack the endogenous purinergic receptor. In these cells, only leukotriene B$_4$ (300 nM) increased calcium levels, while adenosine triphosphate (*ATP*) or ATPγS had no effect

Kennedy 1998). In parallel, we isolated BLT from the porcine leukocytes, and reconstitution experiments were done with several kinds of G-protein in vitro (Igarashi et al. 1999).

## 7.4 Primary Structures of BLT from Various Animals

We have cloned BLT from human, guinea pig and mouse. Table 2 shows the similarities of the sequences of BLTs from these three species (Masuda et al. 1999). Two putative *N*-glycosylation sites and several possible sites of phosphorylation by protein kinase C (PKC) were well conserved among these three species. The homology was higher in the putative transmembrane domains (TMs), especially in TM2, TM3 and TM7. The amino acids in the intracellular loops and tail were also conserved, especially in the third intracellular loop and the proximal portion of the cytoplasmic tail. The third intracellular loop contains a unique cluster of basic amino-acid residues (Fig. 8), suggesting that this region might be involved in G-protein activation and signal transduction. It has only low similarity (less than 20%) with prostaglandin and platelet-activating-factor receptors; instead, it has relatively high similarity (25%–30%) with Gi-coupled chemokine receptors. A recent study has shown that BLT couples with Gi and G16 but not with Gq/11 (Gaudreau et al. 1998).

**Table 2.** Classification of leukotriene (LT) receptors. (Anonymous 1998). For more detailed clinical data, see Brooks and Summers (1996)

| Nomenclature | BLT, Cys-LT1, Cys-LT2 |
|---|---|
| Potency | LTB$_4$≈LTD$_4$, LTC$_4$>LTE$_4$, LTC$_4$>LTD$_4$>LTE$_4$ |
| Antagonist | SB209274: ICI204219 |
| | SC53228: MK476 |
| | ONO4057: SR2640 |
| | CP105696: SKF104353 |
| | CGS25019C: LY170680 |
| | LY293111: ONO1078 |
| Tissue distribution | Leukocytes, spleen, alveolar macrophages, pulmonary vein, thymus, bronchus |
| G-protein effectors | Gq, Gi Gq, Gi, Gq? |
| Chromosome | 14q11.2–12, X |

*BLT*, leukotriene-B$_4$ receptor.

```
                          TM1                              TM2
Human       1 M--NTTSSAA-PPSLGVEFISLLAIILLSVALAVGLPGNSFVVWSILKRMQKRSVTALMVLNLALADIAVLLTAPFFLHF  77
Mouse       1 MAANTTSPAA-PSSPGGMSLSLLPIVLLSVALAVGLPGNSFVVWSILKRMQKRTVTALLVLNLALADLAVLLTAPFFLHF  79
Guinea pig  1 MDRNTTTRAASP-SGSNTFIPLLAMILLSVSMVVGLPGNTFVVWSILKRMRKRSVTALMVLNLALADLAVLLTAPFFLHF  79
              *.***..**.*.....**.******.*.***********.********.*****************

                          TM3                              TM4
Human      78 LAQGTWSFGLAGCRLCHYVCGVSMYASVLLITAMSLDRSLAVARPFVSQKLRTKAMARRVLAGIWVLSFLLATPVLAYRT 157
Mouse      80 LARGTWSFREMGCRLCHYVCGISMYASVLLITIMSLDRSLAVARPFMSQKVRTKAFARWVLAGIWVVSFLLAIPVLVYRT 159
Guinea pig 80 LTWHTWSFKLAGCRLCHYICGVSMYASVLLITAMSLDRSLAVASPFLSQKVRTKTAARWLLVGIWGASFLLATPVLAFRK 159
              *..****..******.***********.************...***.**.*.***.*****.***.

                          TM5                              TM6
Human     158 VVPWKTN-MSLCF-PRYPSEGHRAFHLIFEAVTGFLLPFLAVVASYSDIGRRLQARRFRRSRRTGRLVVLIILTFAAFWL 235
Mouse     160 V-KWN-NR-TLICAPNYPNKEHKVFHLLFEAITGFLLPFLAVVASYSDIGRRLQARRFRRSRRTGRLVVLIILAFAAFWL 236
Guinea pig 160 VVKLT-NETDLCLAV-YPSDRHKAFHLLFEAFTGFVVPFLIVVASYADISRRLRVRRFHRRRRTGRLVVIIILAFAAFWL 237
              *... *  ..**.  .*.**.***.***.*****.***.******.*******..*.****.***.

                          TM7
Human     236 PYHVVNLAEAGRALAGQAAGLGLVGKRLSLARNVLIALAFLSSSVNPVLYACAGGGLLRSAGVGFVAKLLEGTGSEASST 315
Mouse     237 PYHLVNLVEAGRTVAGWDKNSPA-GQRLRLARYVLIALAFLSSSVNPVLYACAGGGLLRSAGVGFVVKLLEGTGSEVSST 315
Guinea pig 238 PYHVVDLVEGSRVLAGTLDQSKQQ---LRNARKVCIALAFLSSSVNPLLYACAGGGLLRSAGVGFVAKLLEATGSEAPST 314
              ***.*.*..*..**........  . ..***.*.************.****.******.****,****.**.

Human     316 RRGGSLGQTARSGPAALEPGPSESL-TASSPLKLN-ELN-                                         352
Mouse     316 RRGGTLVQTPKDTPACPEPGPTDSFMTSS-TI-P--ESSK                                         351
Guinea pig 315 RRGGTLAQTVKGIPMAPEPG--ASG---SLDGLKQSESD-                                         348
              ****.*.**...*...***.  .  . .    .
```

**Fig. 8.** Comparison of the primary structures of leukotriene-B$_4$ receptor (BLT) from human, mouse and guinea pig. BLT is a putative seven-transmembrane (*TM*)-type receptor with two possible *N*-glycosylation sites (amino acids 2 and 164 in the human receptor) and several possible sites of phosphorylation by protein kinase C or protein kinase A. *Asterisks* indicate the amino acid conserved in all species

## 7.5 Chemotactic Activity of LTB4 and BLT

A prominent biological activity of $LTB_4$ is its chemotactic activity (Ford-Hutchinson et al. 1980; Bjork et al. 1982). In vivo and in vitro studies have shown that $LTB_4$ is one of the most potent chemoattractants. Once macrophages and neutrophils are activated, they produce several cytokines and inflammatory mediators, including $LTB_4$, and recruit neutrophils, eosinophils and macrophages into the inflammatory lesions. We were interested in whether CHO cells expressing BLT cause chemotaxis. A filter with 8-µm-diameter pores was coated with fibronectin, and CHO cells were placed in the upper chambers. As shown in Fig. 9, only BLT-carrying cells caused migration toward $LTB_4$ through 8-µm pores (right), whereas vector-transfected cells did not show such a migration. Next, we quantitatively measured chemotactic and chemokinetic activities using a multi-well plate reader. Instead of counting the cells, cells were stained with Diff-quick, and the absorbance at 595 nm was measured (Fig. 10). Both chemotactic activity and chemokinetic activity showed bell-shaped dose responses, and they were completely blocked by pre-treatment of the cells with 100 ng/ml PTX. These data suggest that Gi proteins are required for $LTB_4$-dependent chemotactic and chemokinetic activities. These data show that CHO cells have an intrinsic machinery for cell movement. If proper receptors are expressed, the cells will move toward that ligand. It is important to see the intracellular signaling of BLT toward chemotaxis in CHO cells (dominant negative transfection, antibody injection, etc.) for in vivo use of CHO cells expressing a chemotactic receptor of interest.

LTB4 plays an important role against bacterial infections and invasion of foreign bodies by recruiting leukocytes to inflamed lesions. Overproduction of LTB4 may also cause initiation and progression of several inflammatory disorders. These include arthritis (Griffiths et al. 1995, 1997), asthma (Turner et al. 1996), psoriasis (Iversen et al. 1997) and ischemic or immune nephritis (Rahman et al. 1988; Oberle et al. 1992; Reyes et al. 1992; Noiri et al. 1999). Furthermore, human BLT was reported to act as a co-receptor for macrophage-tropic human immunodeficiency virus (HIV)-1 strains (Owman et al. 1998), as do several chemokine receptors, CCR5 (Alkhaatib et al. 1996; Choe et al. 1996; Deng et al. 1996; Doranz et al. 1996; Dragic et al. 1996; Endres et al. 1996) and CXCR4 (Feng et al. 1996). BLT expressed on human

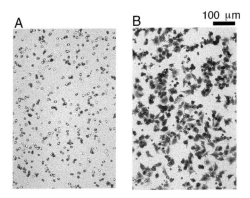

**Fig. 9.** Chemotactic activities of Chinese-hamster ovary (CHO) cells expressing leukotriene-B$_4$ receptor (BLT; Yokomizo et al. 1997). Shown is a photograph of the lower side of the filter in the Boyden-chamber assay. **A** Vector-transfected CHO cells did not cause any migration. Only 8-μM pores are seen. **B** CHO cells expressing human BLT

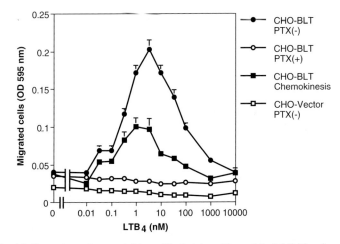

**Fig. 10.** Dose-dependent activities of leukotriene B$_4$ and its inhibition by pertussis toxin (Yokomizo et al. 1997). Shown is the dose–response curve of chemotaxis in Chinese-hamster ovary cells. The numbers of the migrated cells were quantified by measuring the optical density at 595 nm after Diff-quick staining. (mean±SEM, $n=4$). Both chemotaxis and chemokinesis have a maximum at concentrations of 1–10 nM

macrophages can be a target for the inhibition of HIV entry, and various BLT antagonists should be examined for their preventive effects on HIV infection. Thus, development and commercialization of potent and specific BLT antagonists is expected.

**Acknowledgements.** The work was supported, in part, by a grant-in-aid from the Ministry of Education, Science, Sports and Culture of Japan and the Core Research for Evolutional Science and Technology program of the Japan Science and Technology Corporation.

## 7.A Appendix: Cys-LT1 Receptor and Signal Transduction

We found that Cys-LT1 receptor was highly expressed in human THP-1 monocytic leukemia cells. The receptor couples with Gq to increase intracellular Ca (with $EC_{50}$ values of approximately 10 nM) and activates mitogen-activated protein (MAP) kinase ($EC_{50}$=1 nM). It inhibits forskolin-stimulated adenylate cyclase with $IC_{50}$ values of approximately 100 nM. The activation of MAP kinase by Gq protein requires activation PKC and Raf-1 kinase (Hoshino et al. 1998). THP-1 cells showed a chemotactic response toward a low concentration of $LTD_4$ (1–100 nM). The chemotactic responses appear to be related to Gi activation, because treatment of the cells with PTX completely abolished the chemotactic activity. Recently, Dr. Jilly Evans and her collaborators at Merck identified a human Cys-LT1 during the screening of orphan GPCRs with frog-melanophore assays. The receptor was reported in a patent filing as $LTC_4$ responsive and was reported to be $LTD_4$ responsive by another group, though they did not characterize the receptor in detail. Northern blotting and in situ hybridization revealed that the receptor is highly expressed in the human alveolar macrophages and neutrophils. It activates phospholipase C and induces chloride current in *Xenopus* oocyte assays. The binding properties and inhibition by several antagonists suggest that it is Cys-LT1. This was epoch-making research that prompted discovery of various types of related genes, including Cys-LT2, LT-3, etc. Two papers came out (Lynch et al.(1999) Nature 399, 789–793; Sarau et al. (1999) Mol Pharmacol 56, 657–663).

# References

Akbar GKM, Dasari VR, Webb TE, Ayyanathan K, Pillarisetti K, Sandhu AK, Athwal RS, Daniel JL, Ashby B, Barnard EA, Kunapuli SP (1996) Molecular cloning of a novel P2 purinoceptor from human erythroleukemia cells. J Biol Chem 271:18363–18367

Alkhaatib G, Combadiere C, Broder CC, Feng Y, Kennedy PE, Murphy PM, Berger EA (1996) CC CKR5: a RANTES, MIP-1$\alpha$, MIP-1$\beta$ receptor as a fusion cofactor for macrophage-tropic HIV-1. Science 272:1955–1958

Anonymous (1998) 1998 Receptor and ion channel nomenclature. Trends Pharmacol Sci 1998[suppl]:1–98

Bjork J, Hedqvist P, Arfors KE (1982) Increase in vascular permeability induced by leukotriene $B_4$ and the role of polymorphonuclear leukocytes. Inflammation 6:189–200

Bonventre JV, Huang Z, Taheri MR, O'Leary E, Li E, Moskowitz MA, Sapirstein A (1997) Reduced fertility and post-ischaemic brain injury in mice deficient in cytosolic phospholipase A2. Nature 390:622–625

Brooks CD, Summer JB (1996) Modulators of leukotriene biosynthesis and receptor activation. J Med Chem 39:2629–2654

Chen X-S, Sheller R, Johnson N, Func C (1993) Role of leukotriene revealed by targeted disruption of the 5-lipoxygenase gene. Nature 372:179–182

Choe H, Farzan M, Sun Y, Sullivan N, Rollins B, Ponath PD, Wu L, Mackay CR, LaRosa G, Newman W, Gerard N, Gerard C, Sodroski J (1996) The $\beta$-chemokine receptors CCR3 and CCR5 facilitate infection by primary HIV-1 isolates. Cell 85:1135–1148

Deng H, Liu R, Ellmeier W, Choe S, Unutmaz D, Burkhart M, Di Marzio P, Marmon S, Sutton RE, Hill CM, Davis CB, Peiper SC, Schall TJ, Littman DR, Landau NR (1996) Identification of a major co-receptor for primary isolates of HIV-1. Nature 381:661–666

Devchand PR, Keller H, Peters JM, Vazquez M, Gonzalez FJ, Wahli W (1996) The PPAR$\alpha$–leukotriene $B_4$ pathway to inflammation control. Nature 384:39–43

Dixon RA, Jones RE, Diehl RE, Bennett CD, Kargman S, Rouzer CA (1988) Cloning of the cDNA for human 5-lipoxygenase. Proc Natl Acad Sci U S A 85:416-420

Doranz BJ, Rucker J, Yi Y, Smyth RJ, Samson M, Peiper SC, Parmentier M, Collman RG, Doms RW (1996) A dual-tropic primary HIV-1 isolate that uses fusin and the $\beta$-chemokine receptors CKR-5, CKR-3, and CKR-2b as fusion cofactors. Cell 85:1149–1158

Dragic T, Litwin V, Allaway GP, Martin SR, Huang Y, Nagashima KA, Cayanan C, Maddon PJ, Koup RA, Moore JP, Paxton WA (1996) HIV-1 entry into CD4+ cells is mediated by the chemokine receptor CC-CKR-5. Nature 381:667–673

Endres MJ, Clapham PR, Marsh M, Ahuja M, Turner JD, McKnight A, Thomas JF, Stoebenau-Haggarty B, Choe S, Vance PJ, Wells TN, Power CA, Sutterwala SS, Doms RW, Landau NR, Hoxie JA (1996) CD4-independent infection by HIV-2 is mediated by fusin/CXCR4. Cell 87:745–756

Ensor CM, Zhang H, Tai HH (1998) Purification, cDNA cloning and expression of 15-oxoprostaglandin 13-reductase from pig lung. Biochem J 330:103–108

Feng Y, Broder CC, Kennedy PE, Berger EA (1996) HIV-1 entry cofactor: functional cDNA cloning of a seven-transmembrane, G-protein-coupled receptor. Science 272:872–877

Ford-Hutchinson A, Doig MV, Shipley ME, Smith MJ (1980) Leukotriene $B_4$, a potent chemokinetic and aggregating substance released from polymorphonuclear leukocytes. Nature 286:264–265

Gaudreau R, Le Gouill C, Metaoui S, Lemire S, Stankova J, Rola-Pleszczynski M (1998) Signalling through the leukotriene $B_4$ receptor involves both $\alpha i$ and $\alpha 16$, but not $\alpha q$ or $\alpha 11$ G-protein subunits. Biochem J 335:15–18

Griffiths RJ, Pettipher ER, Koch K, Farrell CA, Breslow R, Conklyn MJ, Smith MA, Hackman BC, Wimberly DJ, Milici AJ, McNeish JD (1995) Leukotriene $B_4$ plays a critical role in the progression of collagen-induced arthritis. Proc Natl Acad Sci U S A 92:517–521

Griffiths RJ, Smith MA, Roach ML, Stock JL, Stam EJ, Milici AJ, Scampoli DN, Eskra JD, Byrum RS, Koller BH, McNeish JD (1997) Collagen-induced arthritis is reduced in 5-lipoxygenase-activating protein-deficient mice. J Exp Med 185:1123–1129

Hirabayashi T, Kume K, Hirose K, Yokomizo T, Iino M, Itoh H, Shimizu T (1999) Critical duration of intracellular $Ca^{2+}$ response required for continuous translocation and activation of cytosolic phospholipase $A_2$. J Biol Chem 274:5163–5169

Hoshino M, Izumi T, Shimizu T (1998) Leukotriene $D_4$ activates mitogen-activated protein kinase through a protein kinase $C\alpha$-Raf-1-dependent pathway in human monocytic leukemia THP-1 cells. J Biol Chem 273:4878–4882

Huang WW, Garcia-Zepeda EA, Sauty A, Oettgen HC, Rothenberg ME, Luster AD (1998) Molecular and biological characterization of the murine leukotriene B4 receptor expressed on eosinophils. J Exp Med 188:1063–1074

Igarashi T, Yokomizo T, Tsutsumi O, Taketani Y, Shimizu T, Izumi T (1999) Characterization of the leukotriene $B_4$ receptor in porcine leukocytes. Separation and reconstitution with heterotrimeric GTP-binding proteins. Eur J Biochem 259:419–425

Iversen L, Kragballe K, Ziboh VA (1997) Significance of leukotriene-$A_4$ hydrolase in the pathogenesis of psoriasis. Skin Pharmacol 10:169–177

Kasimir S, Schönfeld W, Hilger RA, König W (1991) Analysis of leukotriene $B_4$ metabolism in human promyelocytic HL-60 cells. Biochem J 279:283-288

Khakh BS, Kennedy C (1998) Adenosine and ATP: progress in their receptor's structures and functions. Trends Pharmacol Sci 19:39-41

Kikuta Y, Kusunose E, Endo K, Yamamoto S, Sogawa K, Fujii KY, Kusunose M (1993) A novel form of cytochrome P-450 family 4 in human polymorphonuclear leukocytes. cDNA cloning and expression of leukotriene $B_4$ ω-hydroxylase. J Biol Chem 268:9376-9380

Kumlin M, Dahlén S-E (1990) Characterization of formation and further metabolism of leukotrienes in the chopped human lung. Biochim Biophys Acta 1044:201-210

Lam BK, Penrose JF, Freeman GJ, Austen KF (1994) Expression cloning of a cDNA for human leukotriene C4 synthase, an integral membrane protein conjugating reduced glutathione to leukotriene $A_4$. Proc Natl Acad Sci U S A 91:7663-7667

Leslie CC (1997) Properties and regulation of cytosolic phospholipase $A_2$. J Biol Chem 272:16709-16712

Masuda H, Yokomizo T, Izumi T, Shimizu T (1999) cDNA cloning and characterization of guinea-pig leukotriene B4 receptor. Biochem J 342:79-85

Miki I, Watanabe T, Nakamura M, Seyama Y, Ui M, Sato F, Shimizu T (1990) Solubilization and characterization of leukotriene B4 receptor-GTP binding protein complex from porcine spleen. Biochem Biophys Res Commun 166 342-348

Minami M, Ohno S, Kawasaki H, Radmark O, Samuellson B, Jornvall H, Shimizu T, Seyama Y, Suzuki K (1987) Molecular cloning of a cDNA coding for human leukotriene $A_4$ hydrolase. J Biol Chem 262:13873-13876

Noiri E, Dickman K, Miller F, Romanov G, Romanov VI, Shaw R, Chambers AF, Rittling SR, Denhardt DT, Goligorsky MS (1999) Reduced tolerance to acute renal ischemia in mice with a targeted disruption of the osteopontin gene. Kidney Int 56:74-82

Oberle GP, Niemeyer J, Thaiss F, Schoeppe W, Stahl RA (1992) Increased oxygen radical and eicosanoid formation in immune-mediated mesangial cell injury. Kidney Int 42:69-74

Ohishi N, Minami M, Kobayashi J, Seyama Y, Hata J, Yotsumoto H, Takaku F, Shimizu T (1990) Immunological quantitation and immunohistochemical localization of leukotriene $A_4$ hydrolase in guinea pig tissues. J Biol Chem 265:7520-7525

Owman C, Nilsson C, Lolait S J (1996) Cloning of cDNA encoding a putative chemoattractant receptor. Genomics 37:187-194

Owman C, Sabirsh A, Boketoft A, Olde B (1997) Leukotriene $B_4$ is the functional ligand binding to and activating the cloned chemoattractant receptor, CMKRL1. Biochem Biophys Res Commun 240:162–166

Owman C, Garzino-Demo A, Cocchi F, Popovic M, Sabirsh A, Gallo RC (1998) The leukotriene $B_4$ receptor functions as a novel type of co-receptor mediating entry of primary HIV-1 isolates into CD4-positive cells. Proc Natl Acad Sci U S A 95:9530–9534

Radmark O, Shimizu T, Jornvall H, Samuelsson B (1984) Leukotriene $A_4$ hydrolase in human leukocytes. Purification and properties. J Biol Chem 259:12339–12345

Rahman MA, Nakazawa M, Emancipator SN, Dunn MJ (1988) Increased leukotriene $B_4$ synthesis in immune injured rat glomeruli. J Clin Invest 81:1945–1952

Reyes AA, Lefkowith J, Pippin J, Klahr S (1992) Role of the 5-lipooxygenase pathway in obstructive nephropathy. Kidney Int 41:100–106

Rouzer CA, Shimizu T, Samuelsson B (1985) On the nature of the 5-lipoxygenase reaction in human leukocytes: characterization of a membrane-associated stimulatory factor. Proc Natl Acad Sci U S A 82:7505-7509

Rouzer CA, Matsumoto T, Samuelsson B (1986) Single protein from human leukocytes possesses 5-lipoxygenase and leukotriene $A_4$ synthase activities. Proc Natl Acad Sci U S A 83:857–861

Samuelsson B, Dahlén SE, Lindgren JÅ, Rouzer CA, Serhan CN (1987) Leukotrienes and lipoxins: structures, biosynthesis, and biological effects. Science 237:1171–1176

Serhan CN, Haeggstrom JZ, Leslie CC (1996) Lipid mediator networks in cell signaling: update and impact of cytokines. FASEB J 10:1147–1158

Shimizu T, Wolfe LS (1990) Arachidonic acid cascade and signal transduction. J Neurochem 55:1–15

Shimizu T, Radmark O, Jornvall H, Samuelsson B (1984) Enzyme with dual activities catalyzes formation of leukotriene $A_4$ from arachidonic acid. Proc Natl Acad Sci U S A 81:689–693

Shimizu T, Izumi T, Seyama Y, Tadokoro K, Radmark O, Samuelsson B (1986) Characterization of leukotriene $A_4$ synthase from murine mast cells: evidence for its identity to arachidonate 5-lipoxygenase. Proc Natl Acad Sci U S A 83:4175–4179

Soberman RJ, Harper TW, Murphy RC, Austen KF (1985) Identification and functional characterization of leukotriene $B_4$ 20-hydroxylase of human polymorphonuclear leukocytes. Proc Natl Acad Sci U S A 82:2292–2295

Turner CR, Breslow R, Conklyn MJ, Andresen CJ, Patterson DK, Lopez AA, Owens B, Lee P, Watson JW, Showell HJ (1996) In vitro and in vivo effects of leukotriene $B_4$ antagonism in a primate model of asthma. J Clin Invest 97:381–387

Ueda N, Kaneko S, Yoshimoto T, Yamamoto S (1986) Purification of arachidonate 5-lipoxygenase from porcine leukocytes and its reactivity with hydroperoxyeicosatetraenoic acids. J Biol Chem 261:7982-7988

Uozumi N, Kume K, Nagase T, Nakatani N, Ishii S, Tashiro F, Komagata Y, Maki K, Ikuta K, Ouchi Y, Miyazaki J-I, Shimizu T (1997) Role of cytosolic phospholipase $A_2$ in allergic response and parturition. Nature 390:618–622

Welsch DJ, Creely DP, Hauser SD, Mathis KJ, Krivi GG Isakson PC (1994) Molecular cloning and expression of human leukotriene-$C_4$ synthase. Proc Natl Acad Sci U S A 91:9745–9749

Yokomizo T, Izumi T, Takahashi T, Kasama T, Kobayashi Y, Sato F, Taketani Y, Shimizu T (1993) Enzymatic inactivation of leukotriene B by a novel enzyme found in the porcine kidney. Purification and properties of leukotriene B 12-hydroxydehydrogenase. J Biol Chem 268:18128–18135

Yokomizo T, Ogawa Y, Uozumi N, Kume K, Izumi T, Shimizu T (1996) cDNA cloning, expression, and mutagenesis study of leukotriene $B_4$ 12-hydroxydehydrogenase. J Biol Chem 271:2844–2850

Yokomizo T, Izumi T, Chang K, Takuwa Y, Shimizu T (1997) A G-protein-coupled receptor for leukotriene $B_4$ that mediates chemotaxis. Nature 387:620–624

# 8 Lipoxins, Aspirin-Triggered 15-epi-Lipoxin Stable Analogs and Their Receptors in Anti-Inflammation: A Window for Therapeutic Opportunity

C. N. Serhan, B. D. Levy, C. B. Clish, K. Gronert, and N. Chiang

| | | |
|---|---|---|
| 8.1 | Introduction | 143 |
| 8.2 | LX Formation: Cell–Cell Interactions and Intravascular LM Generation | 145 |
| 8.3 | Tetraene-Containing LM: Chemical Signals | 150 |
| 8.4 | LX Bioactivities | 154 |
| 8.5 | Cellular and Tissue Sources | 155 |
| 8.6 | LX Stereoselective Bioactions | 158 |
| 8.7 | LX In Vitro Activities | 167 |
| 8.8 | LX and ATL Bioimpacts: "Stop Signals" | 169 |
| 8.9 | Roles in Humans and Disease: Diagnostic Utility? | 174 |
| 8.10 | Summary | 177 |
| References | | 178 |

## 8.1 Introduction

Multicellular host responses to infection, injury or inflammatory stimuli lead to the formation of lipid mediators (LMs) by the host (Cotran et al. 1994). The host response depends on a complex set of events that initiate inflammation and appear to be aimed at resolution. The factors, events and chemical mediators in the complex web that brings about resolution remain to be fully elucidated (Fig. 1). Among the LMs, the lipoxins

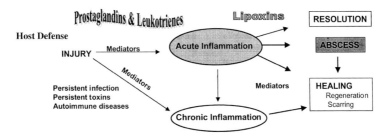

**Fig. 1.** Lipid mediators (LMs) in inflammation. Illustration of the theoretical positions held by LMs in host defense, acute inflammation and resolution. Our working hypothesis states that lipoxins and their aspirin-triggered epimers function in resolution by serving as local mediators in anti-inflammation. See text for discussion

[LXs; i.e., lipoxygenase (LO)-interaction products] are of particular interest, because a body of reported evidence now indicates that, as bioactive LMs, they carry potent anti-inflammatory signals (Serhan 1994; Serhan et al. 1999). It is clear that, during inflammation, LMs play key roles as chemical mediators, initiating and amplifying responses in cell types and tissues and directing the amplitude of the host microenvironmental responses (Fig. 1). Prostaglandins and leukotrienes (LTs) are clearly important in initiating and maintaining these responses (Weissmann 1993; Weissmann et al. 1980). It is now well-appreciated that LXs, first discovered in 1984 in mixed suspensions of human leukocytes incubated with exogenous substrates [i.e., arachidonic acid (C20:4)] or 15-S-hydroperoxyeicosatetraenoic acid [15-S-H(p)ETE; Serhan et al. 1984], are generated in humans by one of at least three biosynthetic routes working independently or in concert (see below) during cell–cell interactions. As "anti-inflammatory" signals in the setting of experimental models, it is highly likely that these conserved structures play pivotal roles in vivo, promoting resolution of inflammatory events. Hence, as endogenous chemical mediators of resolution (of which little is currently known), these compounds [LXs, aspirin-triggered lipoxins (ATLs), their metabolically stable analogs and their cellular targets] are uniquely positioned as therapeutic candidates because they are (1) very potent, (2) amenable to total synthesis strategies, (3) manufacturable and (4) bioavailable. Since this circuit of LX formation and action appears to be of physiological relevance for the resolution of inflammation, therapeutic modalities targeted at this system are likely to have fewer unwanted side effects than other candidates and current anti-inflammatory therapies (Allison et al. 1993). Here, we present an update and our current working models.

## 8.2 LX Formation: Cell–Cell Interactions and Intravascular LM Generation

During LX formation, molecular oxygen is inserted at two sites in C20:4 by distinct LOs that are often segregated into different cell types. Enzymatic oxygenation of C20:4 at two sites is similar to prostaglandin biosynthesis. While LOs have a single oxygenation activity for carbon positions 5, 12 or 15 (nomenclature note: the carbon-1 position is at the carboxylic-acid end of C20:4), cyclooxygenases carry oxygenation activities for both carbon 11 and carbon 15 in the generation of prostanoids (Fig. 2). Thus, by enabling the bi-directional transfer of biosyn-

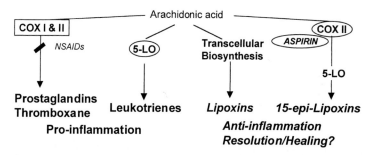

**Fig. 2.** Eicosanoids: new concepts. Unesterified arachidonic acid is rapidly converted to potent bioactive products in human tissues by several pathways that generate functionally distinct classes and series of lipid mediators. Unesterified arachidonic acid is transformed via cyclooxygenase I (*COX I*) or its inducible isoform, COX II, to produce prostaglandins and thromboxane. Leukotrienes are produced via the 5-lipoxygenase pathway, a major route in leukocytes. The lipoxins (LX) are biosynthesized through transcellular interactions, and production of 15-epi-LX is triggered when COX II is acetylated by aspirin. LX and aspirin-triggered lipoxins have potent inhibitory actions and appear to be involved in healing/resolution and anti-inflammation

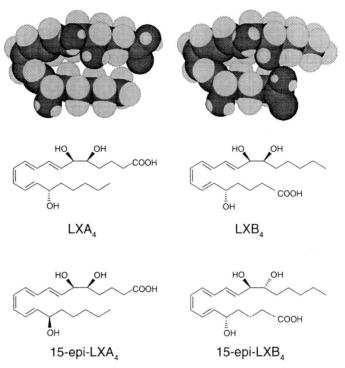

**Fig. 3.** Lipoxin and aspirin-triggered-lipoxin structures. Lipoxin A$_4$ (*LXA$_4$*) is shown as a space-filling model (*upper-left figure*) and as a line drawing (*middle-left figure*). The structure of aspirin-triggered 15-epi-lipoxin A$_4$ (*15-epi-LXA$_4$*), which has its carbon-15 hydroxyl group in the R configuration, is also shown (*lower-left figure*). Lipoxin B$_4$ (*LXB$_4$*) is shown as a space-filling model (*upper-right figure*) and as a line drawing (*middle-right figure*). The structure of aspirin-triggered 15-epi lipoxin B$_4$ (*15-epi-LXB$_4$*) which, like 15-epi-LXA$_4$, also has its carbon-15 hydroxyl group in the R configuration, is also shown (*lower-right figure*)

thetic intermediates between cells, cell–cell interactions with transcellular biosynthesis facilitate the LO catalytic insertions required for LX generation (Serhan 1997). The first report of LX biosynthesis rationalized LX generation by routes involving insertion of molecular oxygen at carbon 15 of C20:4 (predominantly in the S configuration), which implied the involvement of 15-LO (Serhan et al. 1984; Gillmor et al.

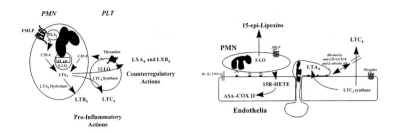

**Fig. 4.** Transcellular routes for lipoxin (LX) and leukotriene (LT) biosynthesis. Leukocyte (polymorphonuclear neutrophil; *PMN*)–platelet (*PLT*) and PMN–endothelia interactions are well-studied models of LX and LT transcellular biosynthesis. Receptor-mediated stimuli (formyl-methionyl-leucyl-phenylalanine for neutrophils and thrombin for PLTs) lead to release of C20:4 from cell membranes for further metabolism [by 5-lipoxygenase (*5-LO*)] to LTA$_4$. This biosynthetic intermediate can be enzymatically converted to either LX (by 12-LO) or LTs (LTB$_4$ by LTA$_4$ hydrolase or LTC$_4$ by LTC$_4$ synthase). During cell–cell interactions, cells can participate in *bidirectional* transcellular biosynthesis; for example, the transfer of PLT C20:4 to neutrophils with subsequent conversion (by 5-LO) to LTA$_4$, which is returned to PLTs for LT or LX generation

1997). Eicosanoid products of 15-LO, 15-S-H(p)ETE or 15-S-hydroxyeicosatetraenoic acid (15-S-HETE) can serve as substrates for 5-LO and lead to the formation of the trihydroxytetraenes lipoxin A$_4$ and lipoxin B$_4$ (LXA$_4$ and LXB$_4$; Fig. 3). These LXs maintain their precursors' alcohol configuration to carry their carbon-15 alcohol in the S configuration.

The second pathway for LX biosynthesis was uncovered for interactions that predominantly occur (in the vasculature) between 5-LO (present in myeloid cells) and 12-LO (in platelets; Fig. 4; Fiore and Serhan 1990). Leukotriene A$_4$ (LTA$_4$), the 5-LO product, is an epoxide intermediate that plays a pivotal role in eicosanoid formation in this scenario. Because more than 50% is released from the cell of origin (Fiore and Serhan 1990), LTA$_4$ serves as an intermediate for both intracellular and transcellular eicosanoid biosynthesis. LTA$_4$ was viewed previously as a highly unstable compound – which it is, as a reactive chemical system; however, within cells and within hydrophobic environments, LTA$_4$'s biological half-life is underappreciated (Fiore and Serhan 1989) and can be prolonged in the presence of phospholipid bilayers. During neutrophil–platelet interaction and co-activation, $LTA_4$ has multiple potential enzymatic and non-enzymatic fates, including (1) conversion to $LXA_4$ and $LXB_4$ by 12-LO, (2) non-enzymatic hydrolysis (which occurs in seconds in aqueous environments), (3) conversion to $LTB_4$ (a potent neutrophil and eosinophil chemoattractant) by $LTA_4$ hydrolase or (4) conversion to $LTC_4$ (a slowly reacting substance produced during anaphylaxis) by $LTC_4$ synthase (Fig. 4). Because $LTB_4$ and $LTC_4$ carry potent pro-inflammatory actions, and because LX inhibits LT-mediated responses in vivo, and the balanced formation of LX and LT is critical for mounting cellular responses.

Of special interest for our understanding of endogenous anti-inflammation mechanisms, drug action and novel structure-based drug design is a third major pathway for LX generation. This pathway was more recently discovered and involves aspirin and the actions of cyclooxygenase II and 5-LO (Clària and Serhan 1995). Many endothelial and epithelial cells express cyclooxygenase II in response to diverse stimuli, such as cytokines, hypoxia and bacterial infection. Aspirin acetylates cyclooxygenase II and triggers its catalytic activity for conversion of C20:4 to 15-R-HETE in lieu of prostanoid biosynthesis (Herschman 1996). During interactions with emigrating peripheral blood leukocytes, 15-R-HETE is released from endothelial and epithelial cells and is transformed into 15-epimer LXs (i.e., ATLs) by leukocyte 5-LO (via transcellular routes; Figs. 3, 5). These 15-R-LXs are even more potent than the 15-S-containing LXs in inhibiting inflammatory responses and cell proliferation, because they resist metabolic inactivation and act within the local microenvironment.

All three of these routes for LX biosynthesis and expression of the $LXA_4$ receptor are subject to modulation by cytokines (Serhan et al. 1996). For example, interleukin 4 (IL-4) and IL-13, which are thought to be negative regulators of the inflammatory response, both increase 15-LO expression (Levy et al. 1993b; Nassar et al. 1994) and activity, thereby enhancing LX formation. Because LXs carry "stop signals" for inflammation, they may serve as local mediators for these anti-inflammatory cytokines. Pro-inflammatory cytokines upregulate 5-LO [granulocyte–macrophage colony-stimulating factor (GM-CSF)] and cyclooxygenase II [IL-1$\beta$, tumor necrosis factor $\alpha$ (TNF$\alpha$)] activities

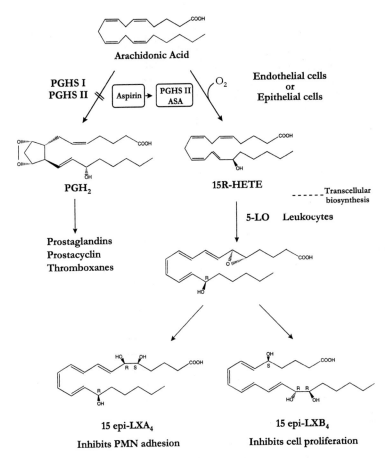

**Fig. 5.** Generation of aspirin-triggered lipoxins. Stimuli (cytokines, hypoxia or lipopolysaccharides) induce the expression of cyclooxygenase II (COX II or PGHS II) in endothelia, epithelia and leukocytes. When aspirin (ASA) is present, COX II is acetylated and no longer converts arachidonic acid (C20:4) to prostaglandins. Instead, it converts C20:4 to 15-R-hydroxyeicosatetraenoic acid, which may then undergo transcellular biosynthesis with neighboring 5-lipoxygenase-bearing cells, such as polymorphonuclear neutrophils, to produce 15-epi-lipoxin A$_4$ and B$_4$ (ASA-triggered lipoxins)

(Serhan et al. 1996), which are crucial to the formation of both LX and ATL. These enzymes are also essential for the generation of LTs and prostaglandins, bioactive lipids that stimulate and act synergistically in experimental inflammation. Therefore, cytokines can orchestrate cellular responses by modulating the balance of anti-inflammatory (LX) and pro-inflammatory (LT) signals (Figs. 1, 2).

## 8.3 Tetraene-Containing LM: Chemical Signals

LXs are trihydroxytetraene-containing eicosanoids (Fig. 3), and this combination of structural features makes them amenable to spectroscopic methods for their characterization and detection. The conjugated tetraene imparts a characteristic ultraviolet (UV)-absorption spectrum on both $LXA_4$ and $LXB_4$, with three intense absorbance bands at $\lambda_{max}^{MeOH}=278$, 300 and 315 nm. A fourth, weaker absorbance band at 270 nm is also present. The most intense absorbance band (at 300 nm), with a molar extinction coefficient of 50,000 $M^{-1}cm^{-1}$, is typically used for UV quantification. In addition, the tetraene structure gives LX fluorescent properties, which may be used in the characterization of their membrane and/or protein binding. At room temperature, this fluorescence has an emission maximum at 410 nm and an excitation maximum at 320 nm (Serhan 1994). A variety of extraction and chromatographic methods may be employed for the purification of LX from complex sample matrices (Levy et al. 1999b). The charge state of the LX carboxylate group may be changed from neutral to negative by altering the pHs of the solutions [and, therefore, the affinity of the compound for reversed-phase stationary phases, such as C18 (i.e., octadecyl) cartridges]. The compound's partition coefficient for liquid phases is also easily modulated in this way. For example, for LX isolation by solid-phase extraction (SPE), samples are typically acidified to protonate the LXs prior to loading onto a C18 cartridge. Following loading, the cartridge may then be washed – first with water to remove more polar moieties, such as salts, and then with hexane to remove materials that are more non-polar. The LXs may then be eluted with a solvent of intermediate polarity, such as methyl formate. Because the tetraene geometry and conformation are fragile, careful handling of LX is required to prevent *cis*-to-*trans* double-bond isomerization. In this regard,

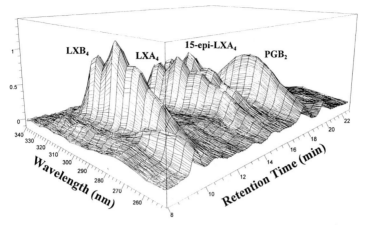

**Fig. 6.** Identification of lipoxins by reversed-phase high-performance liquid chromatography (RP-HPLC) coupled with ultraviolet (UV)-diode-array detection. A RP-HPLC, three-dimensional chromatogram of eicosanoid standards is shown. Eicosanoids (2 ng/each) were analyzed by RP-HPLC with a Hewlett Packard 1100 Series diode-array detector equipped with a binary pump and eluted on a Phenomenex LUNA C18-2 microbore column (150×1 mm, 5 µm) using an isocratic mobile phase composed of methanol/water/acetate (volume ratio: 58/42/0.01) at a flow rate of 0.12 ml/min. Three resolved peaks display UV chromophores with three intense absorbance bands ($\lambda_{max}^{MeOH}$=278, 300 and 315 nm) that are characteristic for the tetraene backbone of lipoxins. These compounds, based on the specific chromophore and retention time, correspond to lipoxin B$_4$ (*LXB$_4$*), LXA$_4$ and 15-epi-LXA$_4$. The fourth peak corresponds to prostaglandin B$_2$, an internal standard, and exhibits a $\lambda_{max}^{MeOH}$=280 nm; therefore, it is easily discernible from lipoxins

LXs are thermally labile and sensitive to light. Therefore, they should be kept cold in an organic solvent, such as methanol, in an environment enriched with nitrogen to prevent non-specific oxygenation.

Other hydroxylated eicosanoids, such as diHETEs and prostaglandins, share some physical properties with LX and, therefore, can co-extract during SPE or liquid–liquid extraction. For further purification, column chromatographic techniques such as gas chromatography (GC) or high-performance liquid chromatography (HPLC) may be employed for analysis. HPLC has advantages, because GC requires functional-group derivatization of LXs for partitioning into the gas phase, while

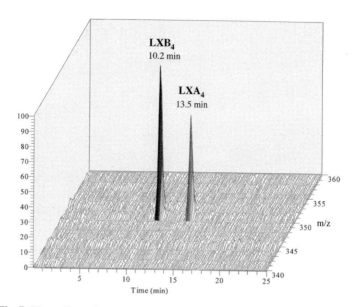

**Fig. 7.** Three-dimensional liquid chromatography–mass spectrometry chromatogram of lipoxins A$_4$ and B$_4$ (*LXA$_4$* and *LXB$_4$*). LXA$_4$ and LXB$_4$ (4.0 ng each) were injected into a high-performance liquid-chromatography system equipped with a LUNA C18-2 column (150×2 mm, Phenomenex, Torrance, Calif.) and were eluted with methanol/water/acetic acid (volumre ratio: 67/33/0.01) directly into the electrospray ionization probe of the mass spectrometer (Finnigan LCQ, Finnigan Corp., San Jose, Calif.). Selected-ion-monitoring mass spectra were recorded in the range m/z=340–360 to detect the molecular anions ([M–H]$^-$) of LXA$_4$ and LXB$_4$, which both have m/z=351.5. LXB$_4$ elutes at 10.2 min while LXA$_4$ elutes at 13.5 min. See Fig. 8 for the product-ion tandem mass spectra of the molecular anions of both LXA$_4$ and LXB$_4$

HPLC requires little or no prior work-up. With a typical reversed-phase C18 column and a methanol/water mobile phase, trihydroxylated compounds, including the LXs, have shorter retention times than dihydroxylated compounds (such as diHETEs) and much shorter retention times than monohydroxylated compounds (such as monoHETEs). A further advantage of HPLC is the facile coupling of UV and mass spectrometric (MS) detection to the apparatus. UV detection with either a photodiode array or a rapid spectrum-scanning detector allows real-time UV profil-

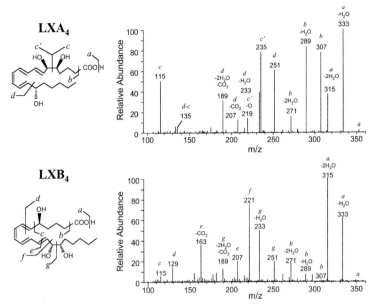

**Fig. 8.** Tandem mass spectrometry (MS/MS) spectra of lipoxins $A_4$ and $B_4$ (*LXA₄* and *LXB₄*). The molecular anions of both LXA₄ and LXB₄ (m/z=351.5; Fig. 7) were fragmented to produce product ion mass spectra (MS/MS). LXA₄ (*upper figure*) shows diagnostic product ions at m/z=333 [351–$H_2O$], 315 [351–2$H_2O$], 307 [351–$CO_2$], 289 [351–($H_2O$+$CO_2$)], 271 [351–(2$H_2O$+$CO_2$)], 251 [351–CHO($CH_2$)$_3$$CH_3$], 235 [351–CHO($CH_2$)$_3$COO$^-$], 233 [351–($H_2O$+CHO($CH_2$)$_4$$CH_3$)], 219 [351–(CHO($CH_2$)$_3$COO$^-$+O)], 207 [351–($CO_2$+CHO($CH_2$)$_4$$CH_3$)], 189 [351–($H_2O$+$CO_2$+CHO($CH_2$)$_4$$CH_3$)], 135 [351–(CHO($CH_2$)$_3$COOH+CHO($CH_2$)$_4$$CH_3$)] and 115 (CHO($CH_2$)$_3$COO$^-$). LXB₄ (*lower figure*) shows diagnostic product ions at m/z=333 [351–$H_2O$], 315 [351–2$H_2O$], 307 [351–$CO_2$], 289 [351–($H_2O$+$CO_2$)], 271 [351–(2$H_2O$+$CO_2$)], 251 [351–CHO($CH_2$)$_4$$CH_3$], 233 [351–($H_2O$+CHO($CH_2$)$_4$$CH_3$)], 221 [351–CHOCHOH($CH_2$)$_4$$CH_3$], 207 [351–($CO_2$+CHO($CH_2$)$_4$$CH_3$)], 189 [351–($H_2O$+$CO_2$+CHO($CH_2$)$_4$$CH_3$)], 163 [351–($CO_2$+$CH_2$COHCHOH($CH_2$)$_4$$CH_3$)], 129 [$CH_3$CO($CH_2$)$_3$COO$^-$] and 115 [CHO($CH_2$)$_3$COO$^-$]

ing of the column eluate and, therefore, allows unambiguous identification of LXs based on both retention times and UV spectra (Fig. 6). In addition, the UV detector may be placed in tandem with a mass spectrometer with an electrospray ionization source. Since the LXs are negatively charged at neutral pH, MS detection in the negative-ion mode is necessary and, depending on the instrument, the LX may be identified by their molecular anions ($[M-H]^-$: m/z=351.5; Fig. 7) or by their characteristic product-ion or tandem MS–MS spectra as they elute from the column (Fig. 8).

By taking advantage of $LXA_4$'s unique overall three-dimensional conformation, an enzyme-linked immunosorbent assay (ELISA) has been developed for rapid detection of $LXA_4$ in multiple samples (Levy et al. 1993a). This ELISA is commercially available, shows no cross-reactivity for 5-S-HETE, 12-S-HETE, 15-S-HETE, $LTB_4$, $LTC_4$, $LTD_4$ or C20:4 and has a $LXA_4$-detection limit of 90 fmol/ml. Furthermore, a selective ELISA for aspirin-triggered 15-epi-$LXA_4$ was also recently developed; it shows little cross-reactivity with native $LXA_4$ or other eicosanoids (Chiang et al. 1998).

## 8.4 LX Bioactivities

### 8.4.1 Pathophysiological Roles

LXs may play a number of physiological and pathophysiological roles (Serhan 1994). Of particular interest, LXs display counter-regulatory action in both in vivo and in vitro models, in sharp contrast to the actions of most other LMs (Serhan 1994). In the nanomolar range, these compounds inhibit neutrophil and eosinophil chemotaxis, neutrophil transmigration across both endothelial and epithelial cells, neutrophil diapedesis from postcapillary venules, and neutrophil entry into inflamed tissues in several animal models. LXs have inhibitory actions on neutrophils, eosinophils and natural killer cells but are potent stimuli of peripheral blood monocyte chemotaxis and adherence (Table 1). While LXs increase monocyte chemotaxis and adherence (Maddox and Serhan 1996), these cells do not degranulate or release reactive oxygen species in response to LX, suggesting that their actions are specific for locomotion and may be related to the recruitment of monocytes to sites of injury

**Table 1.** Leukocyte-selective lipoxin actions

| Human cell type | Lipoxin | Activity | Reference |
|---|---|---|---|
| PMN | LXA$_4$, LXB$_4$ | Blocks emigration in microcirculation | Raud et al. (1991) |
| | | Inhibits chemotaxis | Lee et al. (1989) |
| | | Downregulates CD11/18 | Fiore and Serhan (1995) |
| | | Downregulates intracellular IP$_3$, Ca$^{++}$ | Grandordy et al. (1990) |
| | | Inhibits PMN–endothelial cell and epithelial cell transmigration | Colgan et al. (1993); Papayianni et al. (1996) |
| | 15-epi-LXA$_4$ | Inhibits PMN–endothelial cell interactions | Clària and Serhan (1995) |
| Eosinophils | LXA$_4$ | Inhibits chemotaxis to PAF and fMLP | Soyombo et al. (1994) |
| Monocytes | LXA$_4$ and LXB$_4$ | Stimulates chemotaxis and adherence without degranulation | Maddox and Serhan (1996) |
| Myeloid progenitors | | Stimulates myeloid bone-marrow-derived progenitors | Stenke et al. (1991b) |

*fMLP*, formyl-methionyl-leucyl-phenylalanine; *IP$_3$*, inositol-(1,4,5)-trisphosphate; *LXA$_4$*, lipoxin A$_4$; *LXB$_4$*, lipoxin B$_4$; *PAF*, platelet-activating factor; *PMN*, polymorphonuclear neutrophil.

or inflammation. Hence, LXs are likely to play a role in resolution or repair (Fig. 1) and represent the first chemical mediators in this important event.

### 8.4.2 Ancient Structures of Non-Mammalian Origin

LXA$_4$ and LXB$_4$ are positional isomers (Fig. 3). Their structures are highly conserved and generated across species [from fish (Hill et al. 1999) to humans; see the section on in vitro findings for details]. Structural features of LXA$_4$ are shared by other lipids, including (1) the tetraene, which is also observed in naturally occurring fatty acids, such as parinaric acids (from Fiji nuts; Serhan 1994), (2) carbon positions 1–5, shared with products of 5-LO, and (3) carbon positions 15–20, shared with products of 12-LO, 15-LO and cyclooxygenase.

## 8.5 Cellular and Tissue Sources

Although LX formation by single cell types can be demonstrated in vitro, it appears that transcellular biosynthetic routes are more likely to be a major source of LX during inflammatory events in humans. For

**Table 2.** Stimuli for lipoxin biosynthesis

| Stimulus | Cell–cell interactions | Product |
|---|---|---|
| **Receptor mediated** | | |
| fMLP plus thrombin | Neutrophil–platelet | Lipoxins $A_4$ and $B_4$ |
| GM-CSF, fMLP plus thrombin | Neutrophil–platelet | Lipoxins $A_4$ and $B_4$ |
| PDGF-AB | Neutrophil–platelet | Lipoxins $A_4$ |
| IL-1β, TNFα | Monocyte–astroglial cell | Lipoxins $A_4$ |
| IL-1β, TNFα, LPS, fMLP, thrombin plus aspirin | Neutrophil–endothelial cell | 15-epi-Lipoxin $A_4$ and 15-epi-lipoxin $B_4$ |
| IL-1β plus aspirin | Neutrophil–epithelial cell | 15-epi-Lipoxin $A_4$ and 15-epi-lipoxin $B_4$ |
| **Receptor bypass** | | |
| Divalent cation ionophore (A23187) | Granulocyte–eosinophil | Lipoxins $A_4$ and $B_4$ |
| | Granulocyte–platelet | Lipoxins $A_4$ and $B_4$ |
| | Granulocyte–lung tissue | Lipoxins $A_4$ and $B_4$ |
| | Alveolar macrophage–epithelial cell | Lipoxins $A_4$ and $B_4$ |
| IL-1β, TNFα, LPS, PMA, thrombin plus aspirin | Neutrophil–endothelial cell | 15-epi-Lipoxin $A_4$ and 15-epi-lipoxin $B_4$ |
| Divalent cation ionophore (A23187) | Rainbow trout, Atlantic salmon and carp macrophages (single cell); human alveolar macrophages (single cells) | Lipoxins $A_4$ and $B_4$ |
| GM-CSF, fMLP | Human neutrophils (single cells) | Lipoxins $A_4$ and $B_4$ |

*fMLP*, formyl-methionyl-leucyl-phenylalanine; *GM-CSF*, granulocyte–macrophage colony-stimulating factor; *IL*, interleukin; *LPS*, lipopolysaccharide; *PDGF-AB*, platelet-derived growth factor AB; *PMA*, phorbol myristate acetate; *TNFα*, tumor necrosis factor α.

example, vascular injury or inflammation results in cell–cell interactions between activated peripheral blood leukocytes and platelets, which leads to LX formation via bi-directional transcellular biosynthesis using leukocyte 5-LO and platelet 12-LO (Fiore and Serhan 1990). LX biosynthesis can also occur during cell–cell interactions between infiltrating leukocytes and tissue-resident cells (such as cytokine-primed endothelial cells, which have cyclooxygenase II activity, or epithelial cells which possess 15-LO and cyclooxygenase II activities; Serhan 1997). 5-LO activity was originally demonstrated in neutrophils, but virtually all cells of myeloid lineage are now known to carry 5-LO (Levy et al. 1999b). A variety of non-myeloid cells can also express 5-LO activity, including microglial-specific neurons (including select human brain tumors) and select epithelial cell lines from the intestinal tract (HT-29) and human airway. 15-LO activity is present in high amounts in human

**Table 3.** Human lipoxin (LX) and 15-epi-LX biosynthesis involves cell–cell interactions and cytokine-primed leukocytes in several tissues (Clària and Serhan 1995; Clària et al. 1996; Serhan 1997; Gronert et al. 1998)

| Cell–cell interactions | Transcellular-communication enzymes | Epoxide intermediates | Regulating cytokine |
|---|---|---|---|
| Neutrophil–platelet | 5-LO→12-LO | LTA$_4$ | GM-CSF |
| Neutrophil–mesangial cell | 5-LO→12-LO | LTA$_4$ | Not known |
| Neutrophil–nasal polyp | 15-LO→5-LO | 5(6)Epoxytetraene | |
| Eosinophil-neutrophil | 15-LO→5-LO | 5(6)Epoxytetraene | |
| Eosinophil-neutrophil | 15-LO→5-LO | 5(6)Epoxytetraene | IL-4, IL-13 |
| HIV-infected macrophage–astroglial cell | 5-LO→12-LO | 5(6)Epoxytetraene | |
| In vivo primed neutrophils | 5-LO→15-LO | | |
| Endothelial cell–neutrophil | ASA-COX II→5-LO | 15$R$-hydroxy-5(6)-epoxytetraene | IL-1β, LPS, TNF |
| Airway adenocarcinoma cell (A549)–neutrophil | ASA-COX II→5-LO P450→5-LO | 15$R$-hydroxy-5(6)-epoxytetraene | IL-1β |
| GI epithelial (T84) cells–neutrophils | ASA-COX II | 15$R$-hydroxy-5(6)-epoxytetraene | TNF, IL-1β |

*ASA*, aspirin; *COX II*, cyclooxygenase II; *GM-CSF*, granulocyte–macrophage colony-stimulating factor; *IL*, interleukin; *LO*, lipoxygenase; *LPS*, lipopolysaccharide; *TNF*, tumor necrosis factor.

eosinophils, airway epithelial cells and cytokine-primed monocytes and macrophages; 12-LO is found primarily in human platelets but has also been identified in human intestinal cells from patients with inflammatory bowel disease (Barrett and Bigby 1993).

As might be expected, LX-biosynthetic enzymes are tightly controlled. Table 2 illustrates the receptor-mediated and receptor-bypass stimuli that engage LX biosynthesis by corresponding single cells and cell–cell interactions. Table 3 summarizes the cell–cell interactions and transcellular-communication enzymes involved in LX and 15-epi-LX biosynthesis. Several cytokines (listed in Table 3) can regulate the expression and/or activity of LX or 15-epimer-LX biosynthetic enzymes. In addition, cyclooxygenase II activity can be induced by disease states (including hypoxia and reperfusion injury) as a result of several pro-inflammatory agents, including IL-1β, TNFα and lipopolysaccharides (LPS), and by many growth factors, such as transforming growth factor β, epithelial growth factor, platelet-derived growth factor and fibroblast growth factor (Dubois et al. 1998).

## 8.6 LX Stereoselective Bioactions

### 8.6.1 Ligands for Receptors

The actions of $LXA_4$ in guinea-pig lung strips were proved to be stereospecific; i.e., the 5-S, 6-R orientation of the two hydroxyl groups positioned immediately adjacent to the carboxylic end of the conjugated tetraene is essential for the contractile activity of $LXA_4$, suggesting the presence of specific $LXA_4$-recognition sites. These early structure–activity relationships are important in establishing the bioresponses evoked by LX, but they may not, in view of recent findings, point to contraction of isolated lung strips as a model for target tissues for endogenous LX actions. It is more likely that these compounds act (in the subnanomolar range) to initiate protective actions. In this regard, $LXA_4$ was also shown to display human-leukocyte-selective actions, which implicates them as endogenous "stop signals" (Table 1; Serhan 1994, 1997). In addition, $LXA_4$ stimulates rapid lipid remodeling and release of C20:4 in human polymorphonuclear leukocytes (PMN), which are sensitive to pertussis-toxin (PTX) treatment (Grandordy et al. 1990; Nigam et al. 1990), suggesting the involvement of a G-protein coupled-receptor (GPCR).

The total synthesis and purification of radiolabeled [11,12-$^3$H]-$LXA_4$ (Brezinski and Serhan 1991) enabled the direct characterization of specific $LXA_4$-binding sites on PMN that are likely to mediate its selective bioaction (Fiore et al. 1992). Intact PMN demonstrated specific and reversible [11,12-$^3$H]-$LXA_4$ binding ($K_d \approx 0.5$ nM and $B_{max} \approx 1830$ sites/PMN), which is modulated by guanosine-stable analogs. These $LXA_4$-binding sites are inducible in HL-60 cells exposed to differentiating agents [retinoic acid, dimethyl sulfoxide (DMSO) and phorbol myristate acetate (PMA)] and confer $LXA_4$-induced phospholipase D (PLD) activation in these cells (Fiore et al. 1993). Together, these findings provide further evidence that $LXA_4$ interacts with specific membrane-associated receptors on human leukocytes; these receptors are classical GPCRs. Based on the finding that functional $LXA_4$ receptors are inducible in promyelocytic lineages (HL-60 cells), several putative receptor complementary DNAs (cDNAs, cloned earlier from myeloid lineages and designated orphans) were examined for their ability to bind and signal with $LXA_4$. When transfected into Chinese hamster ovary (CHO) cells, one of the orphans (denoted previously as pINF114) displayed specific [$^3$H]-LXA$_4$ binding with high affinity ($K_d$=1.7 nM) and demonstrated selectivity when compared with LXB$_4$, LTB$_4$, LTD$_4$ and prostaglandin E$_2$ (PGE$_2$; Fiore et al. 1994). In addition, transfected CHO cells were able to transmit signals with LXA$_4$, activating both guanosine triphosphatase (GTPase) and release of C20:4 from membrane phospholipid, indicating that this cDNA encodes a functional receptor for LXA$_4$ in myeloid cells. Later, a mouse LXA$_4$ receptor cDNA was cloned from a spleen cDNA library and was shown to display specific [$^3$H]-LXA$_4$ binding and LXA$_4$-initiated GTPase activity when transfected into CHO cells (Takano et al. 1997). These reported observations for human and mouse LXA$_4$ receptors represented the first cloned LO-derived eicosanoid receptors.

LXA$_4$ receptor (now abbreviated ALXR, consistent with suggested nomenclature; Coleman et al. 1994) cDNA was originally cloned by several individual groups using formyl-methionyl-leucyl-phenylalanine (fMLP)-receptor (FPR) cDNA as a probe under low-stringency hybridization conditions; ALXR demonstrated high sequence homology (~70%) to FPR. Because of this homology, it was named FPRL1 (FPR-like 1; Murphy et al. 1992) or FPRH1 (Bao et al. 1992) and is also known as FPR2 (Ye et al. 1992) or RFP (receptor related to FPR; Perez et al. 1992). This receptor was also cloned by Nomura et al. (1993) from a human monocyte cDNA library and was described as an orphan receptor denoted HM63. These cDNA sequences are available from GenBank (see the section on gene accession numbers). Deduced amino acid sequences reveal that ALXR belongs to the GPCR superfamily characterized by seven putative transmembrane segments (TMSs; Fig. 9) with N-termini on the extracellular side of the membrane and C-termini on the intracellular side (Baldwin 1993).

We found that the overall homology between human and mouse ALXRs is 76% in nucleotide sequence and 73% in deduced amino acid sequence (Takano et al. 1997). An especially high homology was noted in the sixth TMS and the second intracellular loop, suggesting essential roles for these regions in ligand recognition and G-protein coupling (Fig. 10). Molecular evolution via computer-based sequence analysis indicates that ALXR is only distantly related to prostanoid receptors and instead belongs to a rapidly growing cluster of chemoattractive peptide and chemokine receptors exemplified by fMLP, C5a and IL-8 receptors

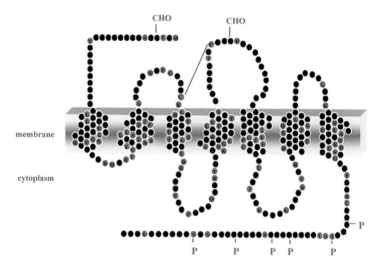

**Fig. 9.** The predicted membrane topology model of lipoxin-$A_4$ receptor (*ALXR*). There is homology between ALXR and leukotroene-$B_4$ receptor (*BLTR*). Deduced amino acid sequence of ALXR demonstrates seven putative transmembrane segments (TMSs) with N-termini on the extracellular side of the membrane and C-termini on the intracellular side. ALXR possesses potential *N*-glycosylation sites (*–CHO*), phosphorylation sites (*–P*) and disulfide linkages (C–C). *Gray circles* indicate the residues conserved between human ALXR (anti-inflammatory) and BLTR (proinflammatory). Unlike prostanoid receptors that use arginine in the seventh TMS as a counterion for ligand binding, neither ALXR nor BLTR possess arginine residues in the seventh TMS as a structural feature

(Toh et al. 1995). Along these lines, a $LTB_4$ receptor (BLTR) was cloned from human HL-60 cells (Yokomizo et al. 1997) and mouse eosinophils (Huang et al. 1998) and was found to share an overall homology of approximately 30% with ALXR in deduced amino acid sequences belonging to this cluster. Of interest, we noted an especially highly homologous region (~46% homology) within the second TMS, where the amino acid sequence LNLALAD is present in both ALXR and BLTR. These findings, taken together, provide further evidence that the origin of receptors of LT and LX is distinct from that of prostanoids and other LMs.

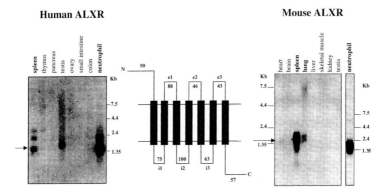

**Fig. 10.** Human and mouse lipoxin-A$_4$ receptor (*ALXR*) homology and tissue distribution. ALXR messenger RNAs (~1.4 kb) in both human and mouse tissues are indicated by *arrows*. *e1–e3* Represent the putative extracellular loops, *T1–T7* are transmembrane segments, and *i1–i3* are intracellular loops for ALXR. The percent homologies between individual domains of human and mouse ALXR in deduced amino acid sequences are indicated by *numbers*. High homology is observed in the second intracellular loop (100%) and the sixth transmembrane segment (97%)

### 8.6.1.1 GenBank Accession Numbers

The cDNA sequences we obtained for human ALXR from THP-1 cells (Maddox et al. 1997) and enterocytes (Gronert et al. 1998) were deposited with GenBank under accession numbers U81501 and AF054013, respectively. They were also deposited earlier (by several individual groups, as discussed above) as FPR homologs and were deposited as orphan receptors under accession numbers X63819 (Perez et al. 1992), M84562 (Murphy et al. 1992), D10922 (Nomura et al. 1993), M88107 (Ye et al. 1992) and M76672 (Bao et al. 1992). Mouse ALXR cDNA was cloned from a mouse-spleen library and is available from GenBank under accession number U78299 (Takano et al. 1997).

Both human (Fiore et al. 1994) and mouse (Takano et al. 1997) ALXR cDNA contain an open reading frame of 1051 nucleotides that encodes a protein of 351 amino acids. Northern-blot analysis (Fig. 10) demonstrated that ALXR messenger RNA (mRNA) is ~1.4 kb in both human and mouse (Takano et al. 1997). Chromosome mapping revealed that the gene encoding human ALXR (Fiore et al. 1994) is located on chromosome 19q (Bao et al. 1992). Hydrophobicity analysis of the deduced amino acid sequence of ALXR revealed seven repeated hydrophobic clusters of 20–25 amino acids interspersed with varying lengths of hydrophilic sequences, which are the common features of GPCR. The seven hydrophobic clusters were proposed to form membrane-spanning α-helices, whereas the hydrophilic segments form loops that project alternately into the extracellular space and the cytoplasm (Baldwin 1993). Several potential post-translational modification sites were observed in the deduced amino acid sequences of both human and mouse ALXR (Fig. 9):

1. Disulfide linkages. Conserved cysteine residues at the first (Cys-99) and second (Cys-176) extracellular loops are found in both human and mouse ALXRs; these residues are proposed to form a disulfide bond to stabilize the tertiary integrity of most GPCRs.
2. $N$-glycosylation. Human ALXR contains two putative $N$-glycosylation sites located at the N-terminus (Asn-4) and second extracellular loop (Asn-179). In mouse ALXR, both putative $N$-glycosylation sites are located at the amino terminus (Asn-4 and Asn-10).
3. Phosphorylation. The C-terminus of mouse ALXR contains nine potential phosphorylation sites (serine and threonine residues), among which six are conserved within the human ALXR.

Intact human PMNs and retinoic-acid-differentiated HL-60 cells demonstrate specific and reversible [$^3$H]-LXA$_4$ binding with $K_d \approx 0.5$ nM and 0.6 nM, respectively (Fiore et al. 1992, 1993). In addition, results from Scatchard analyses indicate that [$^3$H]-LXA$_4$ binds PMN-granule membrane-enriched fractions with a comparable $K_d$ (0.8 nM) but a larger $B_{max}$ ($4.1 \times 10^{-11}$ M) than with plasma-membrane fractions ($K_d$=0.7 nM, $B_{max}$=$2.1 \times 10^{-11}$ M; Fiore et al. 1994). [$^3$H]-LXA$_4$-specific binding is stereoselective, because LTB$_4$, LXB$_4$, 6S-LXA$_4$, 11-$trans$-LXA$_4$ and SKF104353 (a LTD$_4$ antagonist) do not compete for [$^3$H]-LXA$_4$ in human PMNs (Fiore et al. 1992). In addition, among the heteroligands tested on HL-60 cells, only LTC$_4$ (in an excess equal to three times the logarithm of the concentration) competes for [$^3$H]-LXA$_4$-specific binding (Fiore et al. 1993). In contrast, in tissues and cell types other than PMNs, it appears that LXA$_4$ acts to mediate its action as a partial agonist via a subclass of peptido-LT receptors (Badr et al. 1989; Fiore et al. 1992). Both $LTC_4$ and $LXA_4$ were shown to induce contraction of guinea pig lung parenchyma and release of thromboxane $A_2$, which is sensitive to cysteine-LT1-receptor antagonists (Wikstrosm-Jonsson 1998). In addition, $LXA_4$ was shown to block binding of $LTD_4$ to mesangial cells (Badr et al. 1989) and human umbilical-vein endothelial cells (HUVECs; Takano et al. 1997). Furthermore, HUVECs specifically bind [$^3$H]-$LXA_4$ at a $K_d$ of 11 nM, which can be inhibited by $LTD_4$ and SKF104353 (Fiore et al. 1993). Therefore, it appears that $LXA_4$ interacts with at least two classes of cell-surface receptors: one specific for $LXA_4$ on leukocytes, the other shared by $LTD_4$ that is present on HUVEC and mesangial cells. The molecular origin of these $LXA_4/LTD_4$-binding sites is currently of considerable interest.

Human and mouse ALXR cDNA, when transfected into CHO cells, display specific binding to [$^3$H]-$LXA_4$, with $K_d$s of 1.7 nM (Fiore et al. 1994) and 1.5 nM (Takano et al. 1997), respectively. Human-ALXR-transfected CHO cells were also tested for binding with other eicosanoids, including $LXB_4$, $LTD_4$, $LTB_4$ and $PGE_2$. Only $LTD_4$ shows competition with [$^3$H]-$LXA_4$ binding, with a $K_i$ of 80 nM (Fiore et al. 1994). It is of interest to note that, although ALXR shares approximately 70% homology with FPR, ALXR binds [$^3$H]-fMLP with only low affinity ($K_d \approx 5$ μM) and proves to be selective for $LXA_4$ by 3×log(molarity) orders of magnitude (Fiore and Serhan 1995).

In human PMNs, subcellular fractionation revealed that [$^3$H]-$LXA_4$-binding sites are associated with plasma membrane and endoplasmic-reticulum (42.1%), granule (34.5%) and nucleus-enriched fractions (23.3%), a distribution distinct from that of [$^3$H]-$LTB_4$ binding (Fiore et al. 1992). The finding that $LXA_4$ blocks both platelet-activating factor (PAF) and fMLP-stimulated eosinophil chemotaxis (Soyombo et al. 1994) suggests that functional ALXR is also present on eosinophils. In human enterocytes, ALXR is present in crypt and brush- border colonic epithelial cells, as demonstrated by Gronert et al. (1998).

Northern-blot analysis (Fig. 10) of multiple murine tissues demonstrated that, in the absence of challenge to the host, ALXR mRNA is most abundant in neutrophils, spleen and lung, with lesser amounts in heart and liver (Takano et al. 1997). In humans, ALXR mRNA is most abundant in PMNs, followed by spleen, lung, placenta and liver (Fiore et al. 1994; Takano et al. 1997).

In HL-60 cells, retinoic acid, PMA and DMSO, which lead to granulocytic phenotypes, induced a three- to fivefold increase in the expression of ALXR, as monitored by specific [$^3$H]-LXA$_4$ binding (Fiore et al. 1993). Of interest, in human enterocytes, transcription of ALXR is dramatically upregulated by cytokines, of which lymphocyte-derived IL-13 and interferon γ were most potent, followed by IL-4 and IL-6. In addition, IL-1β and LPS also showed moderate induction of ALXR mRNA (Gronert et al. 1998). In view of the cytokine regulation of ALXR, it is likely that the expression of these receptors will change dramatically in disease states; this, in turn, might attenuate mucosal inflammatory and allergic responses.

### 8.6.2 LXA$_4$-Receptor Signaling

In retinoic-acid-differentiated HL-60 cells, LXA$_4$-stimulated PLD activation was shown to be staurosporine sensitive, suggesting the involvement of protein kinase C (PKC) in signal transduction (Fiore et al. 1993). It was also demonstrated that LXA$_4$ blocks LTB$_4$-induced PMN transmigration by inhibiting β2-integrin-dependent PMN adhesion. This modulatory action is partially reversed by prior exposure to genistein, a tyrosine-kinase inhibitor (Papayianni et al. 1996).

The cytoplasmic signaling cascade of ALXR appears to be cell-type specific. For example, in human PMN, LXA$_4$ stimulates rapid lipid remodeling and release of C20:4 via a PTX-sensitive G protein (Nigam et al. 1990) and blocks intracellular generation of inositol-(1,4,5)-trisphosphate (IP$_3$; Grandordy et al. 1990) and Ca$^{2+}$ mobilization (Lee et al. 1989). In contrast, in human monocytes and THP-1 cells, LXA$_4$ triggers intracellular Ca$^{2+}$ release (Romano et al. 1996; Maddox et al. 1997), suggesting a different signaling pathway than in PMNs, despite identical receptor sequences. In addition, distinct signaling in monocytes and PMNs is further evidenced by different responses to LXA$_4$ in these cell types. The characteristics of ALXRs in various cell types are summarized in Table 4. Recently, LXA$_4$ was shown to modulate mitogen-activated-protein-kinase activities on mesangial cells in a PTX-insensitive manner (McMahon et al. 1998), suggesting a novel receptor subtype and/or signaling pathway for ALXR.

**Table 4.** Characterization of human and mouse LXA$_4$ receptor

| Cell type | $K_d$ (nM) | Signal transduction | Kinase associated | Upregulated by |
|---|---|---|---|---|
| Human HL-60 (differentiated by retinoic acid) | 0.6 | PLD activation (lipid remodeling) | Protein kinase C (staurosporin sensitive) | Retinoic acid, DMSO, PMA |
| Human PMN | 0.5 | PLD activation; GTPase activity; C20:4 release; PIPP signal (increased PSDP accumulation with a second signal); no increase of cAMP or proton efflux; very weak [Ca$^{2+}$]$_i$ | Tyrosine kinase (genistein sensitive) | |
| Human PMN (expressed in CHO) | 1.7 | GTPase activity; arachidonic-acid release (PTX sensitive); no increase of cAMP or [Ca$^{2+}$]$_i$ | | |
| Human monocyte | | Increase of [Ca$^{2+}$]$_i$ (PTX sensitive); no increase of cAMP or proton efflux | | |
| Human enterocyte | | No increased proton efflux | | IL-13, IL-4, interferon-γ |
| Human endothelium | 11 | Prostacyclin generation; nitric-oxide generation; no increase of [Ca$^{2+}$]$_i$ or proton efflux | Protein kinase C | |
| Mouse leukocyte (expressed in CHO) | 1.5 | GTPase activity | | |

*cAMP*, cyclic adenosine monophosphate; *CHO*, Chinese hamster ovary; *DMSO*, dimethyl sulfoxide; *GTPase*, guanosine triphosphatase; *IL*, interleukin; *PIPP*, polyisoprenyl phosphate; *PLD*, phospholipase D; *PMA*, phorbol myristate acetate; *PMN*, polymorphonuclear neutrophil; *PSDP*, presqualene diphosphate; *PTX*, pertussis toxin.

In human enterocytes (T84), ALXR activation by a LX analog diminished *Salmonella typhimurium*-induced IL-8 transcription (Gewirtz et al. 1998). The reduction of IL-8 mRNA levels paralleled decreases in IL-8 secretion, suggesting that ALXR's mechanism of action is active at the gene-transcription level.

### 8.6.3 Activating the LXA$_4$ Receptor

ALXR activation in human PMNs evokes inhibition of LTB$_4$- and fMLP-induced PMN adhesion (by downregulating CD11/CD18), chemotaxis, transmigration and degranulation (Serhan 1997). These

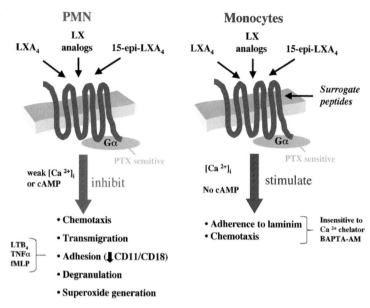

**Fig. 11.** Lipoxin A$_4$ (*LXA$_4$*) and LXA$_4$ receptor (ALXR) evoke different signals and responses in human polymorphonuclear neutrophils (PMNs) and monocytes. ALXR inhibits PMN and stimulates monocyte function via pertussis-toxin-sensitive G proteins (*G$\alpha$*) upon activation by LXA$_4$, 15-epi-LXA$_4$ and LX analogs. In PMN, neither intracellular calcium ($[Ca^{2+}]_i$) nor cyclic adenosine monophosphate were increased in response to lipoxins. In monocytes, LXA$_4$ induced an increase of $[Ca^{2+}]_i$, which is not the second messenger for LXA$_4$-stimulated adherence or chemotaxis, since these responses were unaffected by 1,2-bis(2-aminophenoxy)ethane *N,N,N',N'*-tetraacetic acid acetoxymethyl ester (BAPTA-AM, a Ca$^{2+}$ chelator)

ALXR-mediated, unique biological effects on PMN are summarized in Fig. 11. In human enterocytes, LX analogs inhibit TNF-α-induced IL-8 release (Gronert et al. 1998) and pathogen-induced IL-8 secretion at the mRNA level (Gewirtz et al. 1998). In rabbit trachea, LXA$_4$ stimulates nitric-oxide generation and, therefore, reduces airway smooth-muscle contraction (Tamaoki et al. 1995). In addition, LXA$_4$ analogs inhibit leukocyte rolling and adherence by attenuating P-selectin expression in rat mesenteric microvasculature (Scalia et al. 1997). Recently, it was demonstrated that LX analogs, when applied to mouse ears, dramatically reduce $LTB_4$-induced PMN infiltration and vascular permeability (Takano et al. 1998). Moreover, $LXA_4$ inhibits PMN recruitment to inflamed glomeruli in vivo (Papayianni et al. 1995), further demonstrating the anti-inflammatory actions of LX in vivo.

In vitro, a so-called "receptor knock-out" was carried out using antisense oligonucleotides selected from a region of the ALXR cDNA sequence that had a low homology with the formyl-peptide receptor. In these experiments, retinoic-acid-differentiated HL-60 cells exposed to the selected antisense oligonucleotide specifically diminished [$^3$H]-$LXA_4$ binding and the $LXA_4$-triggered lipid remodeling, which paralleled the loss of ALXR mRNA (Fiore and Serhan 1995). A specific anti-peptide antibody against the second extracellular loop of ALXR was prepared. This antibody blocked specific [$^3$H]-$LXA_4$-binding and $LXA_4$-inhibitory actions on fMLP-stimulated CD11b expression and aggregation on human PMNs (Fiore and Serhan 1995) and inhibited $LXA_4$-triggered intracellular calcium mobilization in human peripheral blood monocytes (Maddox et al. 1997).

## 8.7 LX In Vitro Activities

LXs display a range of highly selective activities with leukocytes. When exposed to $LXA_4$, the PMN $LXA_4$ receptor transduces potent inhibitory action on cellular responses (chemotaxis, adherence, transmigration, granule enzyme release and superoxide-anion generation; Table 1; Serhan 1997). Although many specific signaling events have been elucidated with PMN, our current understanding of the $LXA_4$ receptor's complete intracellular downregulatory signals is of interest. In PMN, LXs do not lead to sustained mobilization of intracellular $Ca^{2+}$, acidification of the intracellular milieu or generation of cyclic adenosine monophosphate but trigger the activation of GTPase, $PLA_2$ and PLD (Serhan 1994). In addition, LXs are not receptor-level antagonists for inflammatory stimuli, such as fMLP or $LTB_4$. For example, LXs inhibit $LTB_4$ responses in PMN, perhaps by uncoupling $LTB_4$-receptor-initiated pro-inflammatory signaling, as evidenced by downregulation of CD11b/CD18, decreased $IP_3$ formation and changes in intracellular PKC distribution (Serhan 1994; Chung-a-on et al. 1996). In addition to their selective actions on PMN, LXs also inhibit eosinophil migration and block natural-killer-cell cytotoxic actions.

In sharp contrast to these downregulatory actions, monocyte adherence and chemotaxis are stimulated by nanomolar concentrations of LX (Maddox and Serhan 1996). These monocyte activities may also be host protective, in view of the important role of monocytes in wound healing and the resolution (Fig. 1) of inflammatory sites (Serhan 1994). In addition to their action on human monocytes, LXs are potent chemoattractants for primordial macrophages from several bony fish, including *Onchorynchus mykiss*, Atlantic salmon and carp (Rowley et al. 1994). These actions on monocytes and macrophages appear to be selective for locomotion, as LXs do not stimulate phagocytosis of microbes or release of reactive oxygen species by these cells. In fact, LX and ATL analogs inhibit degranulation (Gewirtz et al. 1999) and stimulate human macrophages for phagocytosis of apoptotic PMNs (Mitchell et al. 1999). Moreover, LX formation and action on monocytoid cells from rainbow trout (*O. mykiss*) indicate an important role for LXs in leukocyte (particularly monocyte) trafficking, a role that has been conserved during evolution.

In addition to these leukocyte-selective actions, LXs inhibit peptido-LT-induced contraction of isolated vessels and bronchi and myeloid-colony growth (Serhan 1994). $LXA_4$ and $LXB_4$ are potent stimuli (subnanomolar range, with GM-CSF) of the growth of myeloid progenitors (Stenke et al. 1991b).

### 8.7.1 Regulatory Molecules: Inhibitors and Enhancers

Several chemical inhibitors with selective actions on LOs, cyclooxygenase I and II and cytochrome p450 are available. Agents that inhibit 5-LO (zileuton) or peptido-LT-receptors (zafirlukast and montelukast) are now clinically available. These drugs are likely to inhibit LX generation either directly (5-LO inhibition) or indirectly (decreased $LTA_4$ formation with $LTD_4$-receptor antagonism) but have not been directly assayed for their impact on LX biosynthesis. Among the available non-steroidal anti-inflammatory drugs, aspirin is the only known trigger of cyclooxygenase II-mediated 15-R-HETE formation (Clària and Serhan 1995; Herschman 1996).

## 8.8 LX and ATL Bioimpacts: "Stop Signals"

### 8.8.1 Potential Physiological Roles

LXs demonstrate a range of biological actions in vivo that include potent immunologic and hemodynamic regulatory properties (Dahlén and Serhan 1991; Serhan 1994). Their activities are unique and stereospecific for $LXA_4$ and $LXB_4$ structures. In general, LX-receptor activation elicits responses that are both species- and cell-type dependent. Table 6 lists bioactions of LXs, ATLs and their analogs from in vivo experimental models. It is well documented that LXs, in particular $LXA_4$, are potent counter-regulatory signals in vitro for endogenous pro-inflammatory mediators, including $LTB_4$ and PAF, resulting in inhibition of leukocyte-dependent inflammation (Serhan 1994). Leukocyte-selective actions of LX are demonstrated in several in vivo models. In the hamster-cheek pouch, for example, $LXA_4$ markedly inhibits $LTB_4$-induced plasma leakage and leukocyte (predominantly neutrophil) emigration (Dahlén and Serhan 1991).

The cellular and molecular mechanism for LX counter-regulatory actions on PMNs is beginning to unfold, as in vitro data indicate that $LXA_4$ inhibits $LTB_4$- or fMLP-induced CD11b/CD18 upregulation (Fiore and Serhan 1995), a key component in PMN adhesion to both endothelia and epithelia (Serhan 1997). These data are substantiated by rat models that demonstrate the inhibition of P-selectin expression and leukocyte rolling by $LXA_4$ in ileal mesentery (Scalia et al. 1997). Increased leukocyte rolling is considered a pivotal component of the initiation of an inflammatory event. The role of LXs as endogenous "stop signals" for leukocyte recruitment is emphasized by a model of rat glomerulonephritis. PMN exposure to $LXA_4$ dramatically inhibited PMN infiltration into inflamed kidneys (Papayianni et al. 1996). Additional mechanisms are likely, as $LXA_4$ can also counter-regulate $LTB_4$'s action in delayed hypersensitivity in guinea pigs at the level of the $LTB_4$ receptor (Feng et al. 1996). In view of several studies that demonstrate endogenous LX and 15-epi-LX biosynthesis in animal models, it is likely that LXs play an important role in the immune system by acting as in vivo "stop signals" to promote resolution of inflammatory events and protect host tissues.

In addition to demonstrated leukocyte-selective actions, LXs also promote relaxation of smooth muscle (Serhan 1994). LXs have vasodilatory properties and promote vasorelaxation. It is of interest that LXA$_4$ reverses pre-contraction of the pulmonary artery by PGF$_{2\alpha}$ and endothelin-1 (Dahlén and Serhan 1991). Its mechanism of action is endothelium dependent and involves both prostacyclin-dependent and -independent pathways.

LXA$_4$ and LXB$_4$, at nanomolar levels, have been reported to be potent vasoactive eicosanoids by several independent groups (Dahlén and Serhan 1991; Serhan 1994). In some species and tissues, both LXA$_4$ and LXB$_4$ are potent vasodilators while, in other tissues, LXA$_4$ is vasodilatory and LXB$_4$ is a vasoconstrictor. In rat mesenteric-bed and tail arteries, injection of LXA$_4$ induces vasoconstriction. However, in most in vivo models, LXs carry vasodilatory properties at concentrations exceeding those necessary to elicit of leukocyte-directed actions (Clish et al. 1999). LXA$_4$ stimulates rapid vasodilation in hamster cheek pouch and rat renal microcirculation and in cerebral arterioles of newborn pigs. The vasodilatory effect in the hamster cheek pouch and in the cerebral circulation are prostanoid independent. Results from in vitro pharmacologic studies indicate that LXA$_4$ evokes some of these vasoactive properties via a LTD$_4$/LXA$_4$ receptor that has not been identified at the molecular level (vide infra). In rat kidney, LXA$_4$ increased single-nephron glomerular filtration and renal plasma flow rates (Badr et al. 1989). These effects correlate with dilation of afferent arterioles in the glomeruli and appear to counter-regulate the action of LTD$_4$. In vivo administration of LXA$_4$ antagonized the LTD$_4$-induced fall in the rat glomerular-filtration rate but not the renal plasma flow (Badr et al. 1989). Moreover, selective LT-receptor antagonists were shown to block the intrinsic action of LXA$_4$ on mesangial cells (Badr et al. 1989). Taken together, these observations, combined with in vitro binding data (Table 4), are consistent with selective antagonism of LTD$_4$ action at the receptor level in mesangial and vascular endothelial cells. The in vivo models indicate that LXs are potent hemodynamic regulators and, because they are generated during cell–cell interactions, it is likely that these actions are relevant for normal physiological roles.

### 8.8.2 Species Differences

LXs are evolutionarily conserved chemical signals and LMs demonstrated in several species of fish (Serhan 1994), frogs (Gronert et al. 1995) and humans. Their biosynthesis has been demonstrated in numerous animal species. In addition, LX bioactions appear also to be conserved as potent immunologic and vasoactive regulators and are distinct from pro-inflammatory signals, such as LTs and PAF (Serhan 1994, 1997). However, their potencies and actions appear species dependent, as is well known for other LMs and small bioactive molecules.

### 8.8.3 Knockout-Mouse Phenotypes

The role of cell–cell interactions in the biosynthesis of LX at sites of inflammation is emphasized in mouse nephrotoxic serum nephritis. In this model, prominent infiltration of platelets and PMN initiates biosynthesis of large amounts of $LXA_4$. In P-selectin knockout mice, the biosynthesis of this anti-inflammatory eicosanoid is substantially reduced, and leukocyte migration is dysregulated. This deficiency in LX biosynthesis is overcome by transfusing the knockout mice with platelets from wild-type mice that express P-selectin (Mayadas et al. 1996). It is important to note that biosynthesis of LX in the vascular lumen during platelet-PMN interactions places these eicosanoids in a strategic location to inhibit interaction of PMNs and the vascular endothelium.

### 8.8.4 Transgenic Overexpression

In view of 15-LO's role as a key enzyme in LX biosynthesis, it is noteworthy that transgenic overexpression of 15-LO in macrophages protects against atherosclerosis development in rabbits (Shen et al. 1996). It is likely that some of these protective actions are mediated by LX, because IL-4 induction of this enzyme directly initiates their biosynthesis in alveolar macrophages (Levy et al. 1993b).

### 8.8.5 Stable Analogs of LX and ATL

In order to further investigate LX function in inflammatory events, analogs based on the native structures of $LXA_4$ and $LXB_4$ were designed to resist rapid inactivation (Serhan et al. 1995; Takano et al. 1998). Methyl groups were placed on carbon 15 and carbon 5 of $LXA_4$ and $LXB_4$ structures, respectively, to block dehydrogenation by 15-hydroxyprostaglandin dehydrogenase (15-PGDH). 15(R/S)-methyl-$LXA_4$ is a racemic, stable analog of both $LXA_4$ and 15-epi-$LXA_4$. Additional analogs of $LXA_4$ were synthesized with a phenoxy group bonded to carbon 16 and replacing the $\omega$-end of the molecule. This design permits 16-phenoxy-$LXA_4$ to resist potential $\omega$-oxidation and to be protected from dehydrogenation in vivo by the steric hindrance of the bulky aromatic ring. Fluoride was added to the para position of the phenoxy ring to make 16-(para-fluoro)-phenoxy-$LXA_4$ and to hinder non-specific degradation of the phenoxy ring. The aspirin-triggered counterpart of 16-(para-fluoro)-phenoxy-$LXA_4$, 15-epi-16-(para-fluoro)-phenoxy-$LXA_4$, was also synthesized.

The bioactions of LX analogs have been studied in vitro and in vivo and have proved to be potent LX mimetics. Topical application of LX and 15-epi-LX analogs in a mouse-ear model of acute inflammation demonstrated that these analogs are potent inhibitors of $LTB_4$- and PMA-initiated neutrophil recruitment and PMN-mediated vascular injury (Takano et al. 1997, 1998). The cloning of the mouse $LXA_4$ receptor provides direct in vivo evidence of an anti-inflammatory action for both aspirin-triggered $LXA_4$ and $LXA_4$-stable analogs in vivo. As topical agents, these mimetics of endogenous LX proved to be more potent than $LTB_4$-receptor antagonists or the anti-inflammatory steroid dexamethasone (Takano et al. 1998). In addition, in a rat model of endothelial dysfunction characterized by reduced nitric-oxide release and upregulation of adhesion molecules (such as P-selectin), superfusion with LX analogs dramatically inhibited leukocyte rolling (Scalia et al. 1997). Taken together, these data strongly suggest that stable analogs of LX and 15-epi-LX serve as potent, topically active agents that inhibit PMN recruitment and PMN-mediated changes in vascular permeability.

### 8.8.6 Interactions with the Cytokine Network

Even though interaction of cytokine networks with LX bioaction has not been reported in animal models, considerable in vitro data tightly links these two immunoregulatory systems (vide supra). In human enterocytes, $LXA_4$ receptor expression is dramatically upregulated by cytokines that are implicated in the induction of mucosal immune functions. $LXA_4$ and $LXA_4$ analogs inhibit chemokine release at the gene-transcription level in this cell type. Thus, it is likely that LX bioaction in vivo is upregulated by cytokines and that LXs directly modulate the chemokine or cytokine composition at a local inflammatory milieu (Gewirtz et al. 1998; Gronert et al. 1998). Also, GM-CSF enhances LX biosynthesis (Fiore and Serhan 1990) and LXs augment GM-CSF actions on colony formation by myeloid progenitors (Stenke et al. 1991b).

### 8.8.7 Endogenous Inhibitors and Enhancers

Bioactions of LX are amplified when generated locally during cell–cell interactions or multicellular events. In particular, P-selectin-dependent interactions between platelets and leukocytes are important in vivo. In a rat model of glomerulonephritis, in vivo generation of LX during platelet–neutrophil interactions at sites of inflammation was demonstrated (Papayianni et al. 1995). Endogenous generation of LX was inhibited by depleting the animals of either platelets or neutrophils or by treatment with P-selectin antibodies (Papayianni et al. 1995). Induction of acute inflammation in mouse peritoneum and exposure to aspirin initiated ATL-biosynthesis pathways (Chiang et al. 1998). These 15-R-epimers of LX are endogenous mimetics of native 15-S-containing LXs but have enhanced bioactivity, because they resist metabolic inactivation (Serhan 1997). Together, these in vivo results suggest that induction of biosynthetic pathways or cell–cell interactions promotes biosynthesis of LX and 15-epimer LX and, therefore, amplifies the anti-inflammatory action of what appear to be novel LMs of resolution.

## 8.9 Roles in Humans and Disease: Diagnostic Utility?

Although isolated leukocytes and platelets from healthy volunteer blood donors provide a useful model for the study of cell–cell interactions and LX biosynthesis, exogenous stimuli are required to elicit LX formation (Fiore and Serhan 1990). Freshly isolated cells from healthy individuals do not spontaneously generate these compounds. Release of C20:4 from membrane phospholipids is required for LX formation, and no tonic levels of LX are observed in blood or human tissues. Nevertheless, LX formation is routinely observed when cells are exposed to receptor-mediated soluble or phagocytic stimuli (Table 1). Because cells routinely encounter these stimuli, and because LXs carry vasoactive and counter-regulatory actions, $LXA_4$ and $LXB_4$ are likely to have physiologic roles during homeostatic responses, even in the absence of illness. If difficulty is experienced in identifying LXs, it likely reflects the limits of available physical and immunologic detection methods.

Unlike materials from healthy individuals, LXs were identified in vivo during several human illnesses, including asthma, nasal polyps, glomerulonephritis, cirrhotic liver, rheumatoid arthritis, pneumonia, sarcoidosis and after angioplasty-induced rupture of atherosclerotic plaques (Table 5; Serhan 1997). Platelets from patients with chronic myelogenous leukemia exhibit diminished 12-LO activity and are unable to participate in LX generation, which may contribute to the pathophysiology of blast crisis in these patients (Stenke et al. 1994). Unlike circulating PMNs from healthy individuals, which do not generate LX in isolation, PMNs from patients with chronic inflammatory conditions (such as asthma or rheumatoid arthritis) generate LXs (Chavis et al. 1995; Thomas et al. 1995). These disease states likely represent scenarios in which cytokine regulation of key biosynthetic enzymes enables the formation of LXs that are not otherwise observed in healthy individuals.

Currently, only very limited data on the effects of LXs in clinical investigations is available. Nevertheless, in asthmatic patients, inhalation of $LXA_4$ inhibits $LTC_4$-induced airway obstruction and shifts the dose–response curve of airway resistance (Sgaw and $V_{25}$) to the right (Christie et al. 1992). LXs are generated from endogenous sources during provocative challenge in asthma (Levy et al. 1993a), suggesting that they may play roles in modulating airway hyper-responsiveness. In

Table 5. Lipoxins (LXs) and disease

| Organ/system | Findings | Reference |
|---|---|---|
| Hematologic and oncologic | Defect in LX production with cells from chronic-myeloid-leukemia patients in blast crisis | Stenke et al. (1990, 1991a) |
| | LX stimulates the nuclear form of PKC in erythroleukemia cells | Beckman et al. (1992) |
| | Formation of LX by normal human bone-marrow-cell suspensions | Stenke et al. (1991b) |
| | Formation of LX by granulocytes from eosinophilic donors (i.e., having allergic disorders or drug-induced hypereosinophilic syndrome) | Serhan et al. (1987) |
| Vascular | Angioplasty-induced plaque rupture triggers LX formation | Brezinski et al. (1992) |
| Renal | LX triggers renal hemodynamic changes generated in experimental glomerular nephritis | Badr et al. (1987); Papayianni et al. (1993) |
| Dermatologic | $LXA_4$ regulates delayed hypersensitive reactions in skin | Feng et al. (1996) |
| | LX inhibits PMN infiltration and vascular permeability | Takano et al. (1997, 1998) |
| Pulmonary | $LXA_4$ is detected in bronchoalveolar lavage fluids from patients with pulmonary disease and asthma | Lee et al. (1990); Chavis et al. (1995, 1996) |
| | Production of LX by nasal polyps and bronchial tissue | Edenius et al. (1990) |
| | $LXA_4$ inhalation shifts and reduces $LTC_4$-induced contraction in asthmatic patients | Christie et al. (1992) |
| | Aspirin-sensitive asthmatics generate $LXA_4$ on ASA challenge | Levy et al. (1993a) |
| Rheumatoid arthritis | LX levels increase with recovery | Thomas et al. (1995) |

$LXA_4$, lipoxin $A_4$; $LTC_4$, leukotriene $C_4$; *PKC*, protein kinase C.

addition, LXs are formed in human airways in vivo during select inflammatory lung diseases (sarcoidosis, alveolitis and resolving pneumonia; Lee et al. 1990), in cirrhotic ascites (Clària et al. 1998) and intravascularly after percutaneous transluminal angioplasty of atherosclerotic coronary arteries (Brezinski et al. 1992). Together, these limited results provide evidence of a physiological role of LXs in vivo and suggest that prolonging their actions in vivo can have a beneficial effect.

### 8.9.1 An Opportunity for Novel Therapies

Table 6 outlines key findings from preclinical models. Topical application of $LXA_4$, ATL and their stable analogs to mouse-ear skin show vascular-permeability changes and potent inhibition of $LTB_4$- and PMA-driven neutrophil infiltration (Takano et al. 1997, 1998). In addition, delayed hypersensitivity is blocked by $LXA_4$ (Feng et al. 1996). In

**Table 6.** Actions of lipoxins (LXs), aspirin-triggered LX (ATL) and their stable analogs in animal models

| Animal | LX | Model | In vivo effect | Reference |
|---|---|---|---|---|
| Hamster | Native LX | Cheek pouch; vascular network | Vasodilation | Dahlén et al. (1987; 1988) |
|  |  | Cheek pouch, PMN diapedesis | Inhibits $LTB_4$-induced plasma leakage and PMN diapedesis | Hedqvist et al. (1989) |
| Rat | Native LX | Glomerular filtration and renal plasma flow | Antagonizes actions of $LTD_4$ | Badr (1988); Badr et al. (1989) |
|  | LX, ATL-stable analogs | Mesenteric microvasculature | Inhibits leukocyte rolling and P-selectin expression in vascular endothelium | Scalia et al. (1997) |
|  | Native LX | Sleep–wake cycles | Stimulates slow-wave sleep | Kantha et al. (1994) |
| Guinea pig | Native LX | Delayed-type hypersensitivity | Counter-regulates $LTB_4$ action | Feng et al. (1996) |
| Mouse | LX, ATL-stable analogs | Ear (acute inflammation) | Inhibits $LTB_4$-induced PMN infiltration and vascular injury | Takano et al. (1997, 1998); Clish et al. (1999) |
|  | LX, ATL-stable analogs | Air pouch | Blocks PMN exudates in vivo and the cytokine–chemokine axis |  |
| Pigs |  | Cerebral arterioles | Dilation | Busija et al. (1989) |

$LTB_4$, leukotriene $B_4$; $LTD_4$, leukotriene $D_4$; *PMN*, polymorphonuclear neutrophil.

rats, $LXA_4$ also inhibits PMN recruitment to the kidneys of glomerulonephritic animals (Papayianni et al. 1995) and acts as a receptor-level antagonist of $LTD_4$ in kidney mesangial cells, off-setting the physiological contractile actions of $LTD_4$ in these cells (Badr et al. 1989).

### 8.9.2 LX-Inactivation Pathways

LXs are autacoids and, as such, they are rapidly generated in response to stimuli, act within the local microenvironment and are rapidly inactivated. The first step in LX inactivation is dehydrogenation at carbon 15, where the hydroxyl group is oxidized to a ketone group. Interestingly, $LXA_4$ resists ω-oxidation at low concentrations and when incubated with intact human PMN but, at high, non-physiological concentrations, ω-oxidation may be observed (Serhan et al. 1993). This is unlike $LTB_4$, approximately 80% of which is transformed by neutrophils within

10 min under similar conditions. In contrast, when $LXA_4$ is incubated with either differentiated HL-60 cells, intact monocytes or permeabilized monocyte suspensions, 60–80% of the $LXA_4$ is rapidly converted and inactivated within 1 min. The major products are 15-oxo-$LXA_4$, 13,14-dihydro-15-oxo-$LXA_4$ and 13,14-dihydro-$LXA_4$ (Serhan et al. 1993). 15-PGDH catalyzes the conversion of $LXA_4$ to 15-oxo-$LXA_4$ and appears to be the enzyme responsible for catalyzing the first step in the further metabolism of $LXA_4$ (Serhan et al. 1995). This compound is biologically inactive (Serhan et al. 1995) and is further converted to 13,14-dihydro-15-oxo-$LXA_4$ and 13,14-dihydro-$LXA_4$ by reductases that have not been identified. More recently, it has been shown that $LXB_4$ is also dehydrogenated by 15-PGDH at carbon 5 to produce 5-oxo-$LXB_4$; therefore, $LXB_4$ may share a common route of inactivation (Maddox et al. 1998).

## 8.10 Summary

LXs and 15-epimer LXs are generated during cell–cell interactions that occur during multicellular host response to inflammation, tissue injury or host defense. Results indicate that they are present in vivo during human illness and carry predominantly counter-regulatory biological actions opposing the action of well-characterized mediators of inflammation that appear to lead to resolution of the inflammatory response or promotion of repair and wound healing. The first selective receptor of $LXA_4$ was identified by direct ligand binding and was cloned and characterized. Its signaling involves a novel polyisoprenyl-phosphate pathway that directly regulates PLD (Levy et al. 1999a). LX- and 15-epimer-LX-stable analogs that resist metabolic inactivation were designed, synthesized and shown to be potent LX mimetics and novel topically active anti-inflammatory agents in animal models. These new investigational tools enable structure–function studies of LX signal transduction, further elucidation of the role of LX and 15-epimer LX in host responses and exploitation of their potent bioactions in the design of novel pharmacologic agents.

**Acknowledgements.** These studies were supported in part by National Institutes of Health (NIH) grant nos. GM-38765 and P01-DK50305 (C.N. Serhan).

B.D. Levy is a recipient of a mentored clinical scientist development award from the NIH (NHLBI-K08-HL03788); K. Gronert is a recipient of NIH grant no. AI10389, and N. Chiang and C. B. Clish are recipients of postdoctoral fellowships from the Arthritis Foundation.

# References

Allison AC, Lafferty KJ, Fliri H (eds) (1993) Immunosuppressive and anti-inflammatory drugs. The New York Academy of Sciences, New York

Badr KF (1988) The glomerular physiology of lipoxin-A. Adv Exp Med Biol 229:131–136

Badr KF, Serhan CN, Nicolaou KC, Samuelsson B (1987) The action of lipoxin-A on glomerular microcirculatory dynamics in the rat. Biochem Biophys Res Commun 145:408–414

Badr KF, DeBoer DK, Schwartzberg M, Serhan CN (1989) Lipoxin $A_4$ antagonizes cellular and in vivo actions of leukotriene $D_4$ in rat glomerular mesangial cells: evidence for competition at a common receptor. Proc Natl Acad Sci U S A 86:3438–3442

Baldwin JM (1993) The probable arrangement of the helices in G protein-coupled receptors. EMBO J 12:1693–1703

Bao L, Gerard NP, Eddy RL Jr, Shows TB, Gerard C (1992) Mapping of genes for the human C5a receptor (C5AR), human fMLP receptor (FPR), and two fMLP receptor homologue orphan receptors (FPRH1, FPRH2) to chromosome 19. Genomics 13:437–440

Barrett KE, Bigby TD (1993) The intestinal epithelium: a participant in as well as the target of inflammation? Gastroenterology 105:302–303

Beckman BS, Despinasse BP, Spriggs L (1992) Actions of lipoxins $A_4$ and $B_4$ on signal transduction events in Friend erythroleukemia cells. Proc Soc Exp Biol Med 201:169–173

Brezinski DA, Serhan CN (1991) Characterization of lipoxins by combined gas chromatography and electron-capture negative ion chemical ionization mass spectrometry: formation of lipoxin $A_4$ by stimulated human whole blood. Biol Mass Spectrom 20:45–52

Brezinski DA, Nesto RW, Serhan CN (1992) Angioplasty triggers intracoronary leukotrienes and lipoxin $A_4$. Impact of aspirin therapy. Circulation 86:56–63

Busija DW, Armstead W, Leffler CW, Mirro R (1989) Lipoxins $A_4$ and $B_4$ dilate cerebral arterioles of newborn pigs. Am J Physiol 256:H468–H471

Chavis C, Chanez P, Vachier I, Bousquet J, Michel FB, Godard P (1995) 5,15-diHETE and lipoxins generated by neutrophils from endogenous arachidonic acid as asthma biomarkers. Biochem Biophys Res Commun 207:273–279

Chavis C, Vachier I, Chanez P, Bousquet J, Godard P (1996) 5(S),15(S)-Dihydroxyeicosatetraenoic acid and lipoxin generation in human polymorphonuclear cells: dual specificity of 5-lipoxygenase towards endogenous and exogenous precursors. J Exp Med 183:1633–1643

Chiang N, Takano T, Clish CB, Petasis NA, Tai H-H, Serhan CN (1998) Aspirin-triggered 15-epi-lipoxin $A_4$ (ATL) generation by human leukocytes and murine peritonitis exudates: Development of a specific 15-epi-$LXA_4$ ELISA. J Pharmacol Exp Ther 287:779–790

Christie PE, Spur BW, Lee TH (1992) The effects of lipoxin $A_4$ on airway responses in asthmatic subjects. Am Rev Respir Dis 145:1281–1284

Chung-a-on KO, Soyombo O, Spur BW, Lee TH (1996) Stimulation of protein kinase C redistribution and inhibition of leukotriene $B_4$-induced inositol 1,4,5-trisphosphate generation in human neutrophils by lipoxin $A_4$. Br J Pharmacol 117:1334–1340

Clària J, Serhan CN (1995) Aspirin triggers previously undescribed bioactive eicosanoids by human endothelial cell-leukocyte interactions. Proc Natl Acad Sci U S A 92:9475–9479

Clària J, Lee MH, Serhan CN (1996) Aspirin-triggered lipoxins (15-epi-LX) are generated by the human lung adenocarcinoma cell line (A549)-neutrophil interactions and are potent inhibitors of cell proliferation. Mol Med 2:583–596

Clària J, Titos E, Jiménez W, Ros J, Ginès P, Arroyo V, Rivera F, Rodés J (1998) Altered biosynthesis of leukotrienes and lipoxins and host defense disorders in patients with cirrhosis and ascites. Gastroenterology 115:147–156

Clish CB, O'Brien JA, Gronert K, Stahl GL, Petasis NA, Serhan CN (1999) Local and systemic delivery of a stable aspirin-triggered lipoxin prevents neutrophil recruitment in vivo. Proc Natl Acad Sci U S A 96:8247–8252

Coleman RA, Smith WL, Narumiya S (1994) International Union of Pharmacology classification of prostanoid receptors: Properties, distribution, and structure of the receptors and their subtypes. Pharmacol Rev 46:205–229

Colgan SP, Serhan CN, Parkos CA, Delp-Archer C, Madara JL (1993) Lipoxin $A_4$ modulates transmigration of human neutrophils across intestinal epithelial monolayers. J Clin Invest 92:75–82

Cotran RS, Kumar V, Robbins SL (1994) Robbins pathologic basis of disease. Saunders, Philadelphia

Dahlén S-E, Serhan CN (1991) Lipoxins: bioactive lipoxygenase interaction products. In: Wong A, Crooke ST (eds) Lipoxygenases and their products. Academic, San Diego, p 235

Dahlén SE, Raud J, Serhan CN, Björk J, Samuelsson B (1987) Biological activities of lipoxin A include lung strip contraction and dilation of arterioles in vivo. Acta Physiol Scand 130:643–647

Dahlén SE, Franzén L, Raud J, Serhan CN, Westlund P, Wikström E, Björck T, Matsuda H, Webber SE, Veale CA, Puustinen T, Haeggström J, Nicolaou KC, Samuelsson B (1988) Actions of lipoxin $A_4$ and related compounds in smooth muscle preparations and on the microcirculation in vivo. Adv Exp Med Biol 229:107–130

Dubois RN, Abramson SB, Crofford L, Gupta RA, Simon LS, Van de Putte LBA, Lipsky PE (1998) Cyclooxygenase in biology and disease. FASEB J 12:1063–1073

Edenius C, Kumlin M, Björk T, Anggard A, Lindgren JA (1990) Lipoxin formation in human nasal polyps and bronchial tissue. FEBS Lett 272:25–28

Feng Z, Godfrey HP, Mandy S, Strudwick S, Lin K-T, Heilman E, Wong PY-K (1996) Leukotriene $B_4$ modulates in vivo expression of delayed-type hypersensitivity by a receptor-mediated mechanism: regulation by lipoxin $A_4$. J Pharmacol Exp Ther 278:950–956

Fiore S, Serhan CN (1989) Phospholipid bilayers enhance the stability of leukotriene $A_4$ and epoxytetraenes: stabilization of eicosanoids by liposomes. Biochem Biophys Res Commun 159:477–481

Fiore S, Serhan CN (1990) Formation of lipoxins and leukotrienes during receptor-mediated interactions of human platelets and recombinant human granulocyte/macrophage colony-stimulating factor-primed neutrophils. J Exp Med 172:1451–1457

Fiore S, Serhan CN (1995) Lipoxin $A_4$ receptor activation is distinct from that of the formyl peptide receptor in myeloid cells: inhibition of CD11/18 expression by lipoxin $A_4$–lipoxin $A_4$ receptor interaction. Biochemistry 34:16678–16686

Fiore S, Ryeom SW, Weller PF, Serhan CN (1992) Lipoxin recognition sites. Specific binding of labeled lipoxin $A_4$ with human neutrophils. J Biol Chem 267:16168–16176

Fiore S, Romano M, Reardon EM, Serhan CN (1993) Induction of functional lipoxin $A_4$ receptors in HL-60 cells. Blood 81:3395–3403

Fiore S, Maddox JF, Perez HD, Serhan CN (1994) Identification of a human cDNA encoding a functional high affinity lipoxin $A_4$ receptor. J Exp Med 180:253–260

Gewirtz AT, McCormick B, Neish AS, Petasis NA, Gronert K, Serhan CN, Madara JL (1998) Pathogen-induced chemokine secretion from model intestinal epithelium is inhibited by lipoxin $A_4$ analogs. J Clin Invest 101:1860–1869

Gewirtz AT, Fokin VV, Petasis NA, Serhan CN, Madara JL (1999) $LXA_4$, aspirin-triggered 15-epi-$LXA_4$, and their analogs selectively downregulate PMN azurophilic degranulation. Am J Physiol 276:C988–C994

Gillmor SA, Villasenor A, Fletterick R, Sigal E, Browner MF (1997) The structure of mammalian 15-lipoxygenase reveals similarity to the lipases and the determinants of substrate specificity. Nat Struct Biol 4:1003–1009

Grandordy BM, Lacroix H, Mavoungou E, Krilis S, Crea AE, Spur BW, Lee TH (1990) Lipoxin $A_4$ inhibits phosphoinositide hydrolysis in human neutrophils. Biochem Biophys Res Commun 167:1022–1029

Gronert K, Virk SM, Herman CA (1995) Endogenous sulfidopeptide leukotriene synthesis and 12-lipoxygenase activity in bullfrog (*Rana catesbeiana*) erythrocytes. Biochim Biophys Acta 1255:311–319

Gronert K, Gewirtz A, Madara JL, Serhan CN (1998) Identification of a human enterocyte lipoxin $A_4$ receptor that is regulated by IL-13 and IFN-$\gamma$ and inhibits TNF-$\alpha$-induced IL-8 release. J Exp Med 187:1285–1294

Hedqvist P, Raud J, Palmertz U, Haeggström J, Nicolaou KC, Dahlén SE (1989) Lipoxin $A_4$ inhibits leukotriene $B_4$-induced inflammation in the hamster cheek pouch. Acta Physiol Scand 137:571–572

Herschman HR (1996) Prostaglandin synthase 2. Biochim Biophys Acta 1299:125–140

Hill DJ, Griffiths DH, Rowley AF (1999) Trout thrombocytes contain 12- but not 5-lipoxygenase activity. Biochim Biophys Acta 1437:63–70

Huang W-W, Garcia-Zepeda EA, Sauty A, Oettgen HC, Rothenberg ME, Luster AD (1998) Molecular and biological characterization of the murine leukotriene $B_4$ receptor expressed on eosinophils. J Exp Med 188:1063–1074

Kantha SS, Matsumura H, Kubo E, Kawase K, Takahata R, Serhan CN, Hayaishi O (1994) Effect of prostaglandin $D_2$, lipoxins and leukotrienes on sleep and brain temperature of rats. Prostaglandins Leukot Essent Fatty Acids 51:87–93

Lee TH, Horton CE, Kyan-Aung U, Haskard D, Crea AE, Spur BW (1989) Lipoxin $A_4$ and lipoxin $B_4$ inhibit chemotactic responses of human neutrophils stimulated by leukotriene $B_4$ and $N$-formyl-l-methionyl-l-leucyl-l-phenylalanine. Clin Sci 77:195–203

Lee TH, Crea AE, Gant V, Spur BW, Marron BE, Nicolaou KC, Reardon E, Brezinski M, Serhan CN (1990) Identification of lipoxin $A_4$ and its relationship to the sulfidopeptide leukotrienes $C_4$, $D_4$, and $E_4$ in the bronchoalveolar lavage fluids obtained from patients with selected pulmonary diseases. Am Rev Respir Dis 141:1453–1458

Levy BD, Bertram S, Tai HH, Israel E, Fischer A, Drazen JM, Serhan CN (1993a) Agonist-induced lipoxin $A_4$ generation: detection by a novel lipoxin $A_4$-ELISA. Lipids 28:1047–1053

Levy BD, Romano M, Chapman HA, Reilly JJ, Drazen J, Serhan CN (1993b) Human alveolar macrophages have 15-lipoxygenase and generate 15(S)-hydroxy-5,8,11-*cis*-13-*trans*-eicosatetraenoic acid and lipoxins. J Clin Invest 92:1572–1579

Levy BD, Fokin VV, Clark JM, Wakelam MJO, Petasis NA, Serhan CN (1999a) Polyisoprenyl phosphate (PIPP) signaling regulates phospholipase D activity: a "stop" signaling switch for aspirin-triggered lipoxin $A_4$. FASEB J 13:903–911

Levy BD, Gronert K, Clish C, Serhan CN (1999b) Leukotriene and lipoxin biosynthesis. In: Laychock S, Rubin RP (eds) Lipid second messengers. CRC, Boca Raton, p 83

Maddox JF, Serhan CN (1996) Lipoxin $A_4$ and $B_4$ are potent stimuli for human monocyte migration and adhesion: selective inactivation by dehydrogenation and reduction. J Exp Med 183:137–146

Maddox JF, Hachicha M, Takano T, Petasis NA, Fokin VV, Serhan CN (1997) Lipoxin $A_4$ stable analogs are potent mimetics that stimulate human monocytes and THP-1 cells via a G-protein linked lipoxin $A_4$ receptor. J Biol Chem 272:6972–6978

Maddox JF, Colgan SP, Clish CB, Petasis NA, Fokin VV, Serhan CN (1998) Lipoxin $B_4$ regulates human monocyte/neutrophil adherence and motility: design of stable lipoxin $B_4$ analogs with increased biologic activity. FASEB J 12:487–494

Mayadas TN, Mendrick DL, Brady HR, Tang T, Papayianni A, Assmann KJM, Wagner DD, Hynes RO, Cotran RS (1996) Acute passive anti-glomerular basement membrane nephritis in P-selectin-deficient mice. Kidney Int 49:1342–1349

McMahon B, McPhilips F, Fanning A, Brady HR, Godson C (1998) Modulation of mesangial cell MAP kinase activities by leukotriene $D_4$ and lipoxin $A_4$. J Am Soc Nephrol 9:355A

Mitchell S, Harvey K, Fokin VV, Petasis NA, Godson C, Brady HR (1999) Lipoxins stimulate macrophage phagocytosis of apoptotic neutrophils (abstract). FASEB J 13:A557

Murphy PM, Ozcelik T, Kenney RT, Tiffany HL, McDermott D (1992) A structural homologue of the $N$-formyl peptide receptor: characterization and chromosomal mapping of a peptide chemoattractant receptor gene family. J Biol Chem 267:7637–7643

Nassar GM, Morrow JD, Roberts LJ, II, Lakkis FG, Badr KF (1994) Induction of 15-lipoxygenase by interleukin-13 in human blood monocytes. J Biol Chem 269:27631–27634

Nigam S, Fiore S, Luscinskas FW, Serhan CN (1990) Lipoxin $A_4$ and lipoxin $B_4$ stimulate the release but not the oxygenation of arachidonic acid in human neutrophils: dissociation between lipid remodeling and adhesion. J Cell Physiol 143:512–523

Nomura H, Nielsen BW, Matsushima K (1993) Molecular cloning of cDNAs encoding a LD78 receptor and putative leukocyte chemotactic peptide receptors. Int Immunol 5:1239

Papayianni A, Takata S, Serhan CN, Brady HR (1993) Counter regulatory actions of leukotrienes (LT) and lipoxins (LX) on P-selectin expression on human endothelial cells and neutrophil-endothelial cell adhesion (abstract). J Am Soc Nephrol 4:627

Papayianni A, Serhan CN, Phillips ML, Rennke HG, Brady HR (1995) Transcellular biosynthesis of lipoxin $A_4$ during adhesion of platelets and neutrophils in experimental immune complex glomerulonephritis. Kidney Int 47:1295–1302

Papayianni A, Serhan CN, Brady HR (1996) Lipoxin $A_4$ and $B_4$ inhibit leukotriene-stimulated interactions of human neutrophils and endothelial cells. J Immunol 156:2264–2272

Perez HD, Holmes R, Kelly E, McClary J, Andrews WH (1992) Cloning of a cDNA encoding a receptor related to the formyl peptide receptor of human neutrophils. Gene 118:303–304

Raud J, Palmertz U, Dahlén SE, Hedqvist P (1991) Lipoxins inhibit microvascular inflammatory actions of leukotriene $B_4$. Adv Exp Med Biol 314:185–192

Romano M, Maddox JF, Serhan CN (1996) Activation of human monocytes and the acute monocytic leukemia cell line (THP-1) by lipoxins involves unique signaling pathways for lipoxin $A_4$ versus lipoxin $B_4$. J Immunol 157:2149–2154

Rowley AF, Lloyd-Evans P, Barrow SE, Serhan CN (1994) Lipoxin biosynthesis by trout macrophages involves the formation of epoxide intermediates. Biochemistry 33:856–863

Scalia R, Gefen J, Petasis NA, Serhan CN, Lefer AM (1997) Lipoxin $A_4$ stable analogs inhibit leukocyte rolling and adherence in the rat mesenteric microvasculature: role of P-selectin. Proc Natl Acad Sci U S A 94:9967–9972

Serhan CN (1994) Lipoxin biosynthesis and its impact in inflammatory and vascular events. Biochim Biophys Acta 1212:1–25

Serhan CN (1997) Lipoxins and novel aspirin-triggered 15-epi-lipoxins (ATL): a jungle of cell-cell interactions or a therapeutic opportunity? Prostaglandins 53:107–137

Serhan CN, Hamberg M, Samuelsson B (1984) Lipoxins: novel series of biologically active compounds formed from arachidonic acid in human leukocytes. Proc Natl Acad Sci U S A 81:5335–5339

Serhan CN, Hirsch U, Palmblad J, Samuelsson B (1987) Formation of lipoxin A by granulocytes from eosinophilic donors. FEBS Lett 217:242–246

Serhan CN, Fiore S, Brezinski DA, Lynch S (1993) Lipoxin $A_4$ metabolism by differentiated HL-60 cells and human monocytes: conversion to novel 15-oxo and dihydro products. Biochemistry 32:6313–6319

Serhan CN, Maddox JF, Petasis NA, Akritopoulou-Zanze I, Papayianni A, Brady HR, Colgan SP, Madara JL (1995) Design of lipoxin $A_4$ stable analogs that block transmigration and adhesion of human neutrophils. Biochemistry 34:14609–14615

Serhan CN, Haeggström JZ, Leslie CC (1996) Lipid mediator networks in cell signaling: update and impact of cytokines. FASEB J 10:1147–1158

Serhan CN, Takano T, Gronert K, Chiang N, Clish CB (1999) Lipoxin and aspirin-triggered 15-epi-lipoxin cellular interactions: anti-inflammatory lipid mediators. Clin Chem Lab Med 37:299–309

Shen J, Herderick E, Cornhill JF, Zsigmond E, Kim H-S, Kühn H, Guevara NV, Chan L (1996) Macrophage-mediated 15-lipoxygenase expression protects against atherosclerosis development. J Clin Invest 98:2201–2208

Soyombo O, Spur BW, Lee TH (1994) Effects of lipoxin $A_4$ on chemotaxis and degranulation of human eosinophils stimulated by platelet-activating factor and $N$-formyl-l-methionyl-l-leucyl-l-phenylalanine. Allergy 49:230–234

Stenke L, Näsman-Glaser B, Edenius C, Samuelsson J, Palmblad J, Lindgren JÅ (1990) Lipoxygenase products in myeloproliferative disorders: increased leukotriene $C_4$ and decreased lipoxin formation in chronic myeloid leukemia. Adv Prostaglandin Thromboxane Leukot Res 21B:883–886

Stenke L, Edenius C, Samuelsson J, Lindgren JA (1991a) Deficient lipoxin synthesis: a novel platelet dysfunction in myeloproliferative disorders with special reference to blastic crisis of chronic myelogenous leukemia. Blood 78:2989–2995

Stenke L, Mansour M, Edenius C, Reizenstein P, Lindgren JA (1991b) Formation and proliferative effects of lipoxins in human bone marrow. Biochem Biophys Res Commun 180:255–261

Stenke L, Reizenstein P, Lindgren JA (1994) Leukotrienes and lipoxins–new potential performers in the regulation of human myelopoiesis. Leuk Res 18:727–732

Takano T, Fiore S, Maddox JF, Brady HR, Petasis NA, Serhan CN (1997) Aspirin-triggered 15-epi-lipoxin $A_4$ and $LXA_4$ stable analogs are potent inhibitors of acute inflammation: Evidence for anti-inflammatory receptors. J Exp Med 185:1693–1704

Takano T, Clish CB, Gronert K, Petasis N, Serhan CN (1998) Neutrophil-mediated changes in vascular permeability are inhibited by topical application of aspirin-triggered 15-epi-lipoxin $A_4$ and novel lipoxin $B_4$ stable analogues. J Clin Invest 101:819–826

Tamaoki J, Tagaya E, Yamawaki I, Konno K (1995) Lipoxin $A_4$ inhibits cholinergic neurotransmission through nitric oxide generation in the rabbit trachea. Eur J Pharmacol 287:233–238

Thomas E, Leroux JL, Blotman F, Chavis C (1995) Conversion of endogenous arachidonic acid to 5,15-diHETE and lipoxins by polymorphonuclear cells from patients with rheumatoid arthritis. Inflamm Res 44:121–124

Toh H, Ichikawa A, Narumiya S (1995) Molecular evolution of receptors for eicosanoids. FEBS Lett 361:17–21

Weissmann G (1993) Prostaglandins as modulators rather than mediators of inflammation. J Lipid Mediat Cell Signal 6:275–286

Weissmann G, Smolen JE, Korchak HM (1980) Release of inflammatory mediators from stimulated neutrophils. N Engl J Med 303:27–34

Wikstrosm-Jonsson E (1998). Functional characterization of receptors for cysteinyl leukotrienes in muscle (doctoral thesis). Karolinka Institute, Stockholm

Ye RD, Cavanagh SL, Quehenberger O, Prossnitz ER, Cocrane C (1992) Isolation of a cDNA that encodes a novel granulocyte $N$-formyl peptide receptor. Biochem Biophys Res Commun 184:582–589

Yokomizo T, Izumi T, Chang K, Takuwa T, Shimizu T (1997) A G-protein-coupled receptor for leukotriene $B_4$ that mediates chemotaxis. Nature 387:620–624

# 9 Lipoxin-Stable Analogs: Potential Therapeutic Downregulators of Intestinal Inflammation

A. T. Gewirtz and J. L. Madara

| | | |
|---|---|---|
| 9.1 | Intestinal Diseases Characterized by Active Inflammation | 187 |
| 9.2 | Pathobiology of Active Intestinal Inflammation | 188 |
| 9.3 | Molecular Basis of Immune Inflammatory Response | 189 |
| 9.4 | Lipoxins are Natural Endogenously Biosynthesized Eicosanoids with Anti-Inflammatory Activity | 190 |
| 9.5 | Results from both In Vitro and In Vivo Studies Indicate that Lipoxin-Stable Analogs Likely Have Therapeutic Potential for Intestinal Diseases Characterized by Active Inflammation | 192 |
| 9.6 | Why $LXA_4$ Analogs, Unlike Non-Steroidal Anti-Inflammatory Drugs, are Unlikely to Exacerbate Intestinal Disease | 194 |
| 9.7 | Conclusions | 196 |
| References | | 196 |

## 9.1 Intestinal Diseases Characterized by Active Inflammation

A number of intestinal diseases are characterized by active intestinal inflammation. These diseases include pathogen-induced disorders (salmonellosis, shigellosis) and chronic inflammatory diseases of the intestine (i.e., Crohn's disease and ulcerative colitis) collectively referred to as inflammatory bowel diseases (IBDs; Yamada et al. 1995). The symptoms (diarrhea, cramping) associated with active intestinal inflammation can be quite debilitating and dangerous (potentially causing severe dehydration). Furthermore, those afflicted with IBD (particularly ulcerative colitis) are at increased risk of colon cancer. While pathogen-induced intestinal disorders can potentially be avoided and are generally self resolving, there is a clear need for therapeutic intervention in IBD. Unfortunately, the repertoire of agents currently available for treatment of intestinal inflammation is most unsatisfactory.

## 9.2 Pathobiology of Active Intestinal Inflammation

The histopathology of active flare-ups of chronic inflammatory intestinal diseases is very similar to the histopathology of the acute disease induced by enteric pathogens (Yardley 1986). In either case, the hallmark of active disease is the presence of polymorphonuclear leukocytes (neutrophils; PMNs) that have migrated to the intestinal lumen. This neutrophil movement is not thought to merely be the defining indicator of active intestinal inflammation but appears to play a major role in causing the sequellae associated with this disease state. Neutrophils that have transmigrated to the intestinal lumen can induce the epithelial cells lining the intestine to secrete chloride ions into the lumen (Madara et al. 1991, 1992, 1993). This chloride secretion drives water movement, which is the basis for the secretory diarrhea that accompanies intestinal inflammation. Furthermore, transmigrating neutrophils impale epithelial tight junctions, thus disrupting this tissue's barrier function (Madara 1989), allowing access of the lumenal contents to subepithelial cells that are not used to bathing in this noxious milieu (Madara 1989). The response of these cells (especially lamina propria macrophages) to exposure to the toxins and microbes of the intestinal lumen will result in further promotion of the inflammatory state (McCormick et al. 1998a). Additionally, the infiltrating PMNs secrete oxidants and proteases that can further damage the epithelia and induce secretion of cytokines/chemokines that will also exacerbate the inflammatory state. While this immune inflammatory response may play a role in host defense against some lumenal pathogens, it is not yet clear why responses very similar to this occur in the absence of any known pathogens in persons afflicted with IBD.

## 9.3 Molecular Basis of Immune Inflammatory Response

While enteric pathogens are clearly able to trigger activation of a mucosal inflammatory response in healthy individuals, a provoking agonist(s) that initiates the active flare-ups in IBD patients has not yet been identified. It has been suggested that normal gut microflora might serve as a trigger because, in mouse models of IBD, development of disease is dependent on the presence of such organisms (Baumgart et al. 1998). The primary host interface with (and barrier against) enteric microorganisms are the epithelial cells that line the intestinal tract. Recent studies have shown that the intestinal epithelium is not a passive barrier but plays an active role in orchestrating the immune inflammatory response. In response to enteric pathogens, the epithelium secretes cytokines/chemokines that recruit and activate immune cells. Epithelial-cell-secreted cytokines include pleiotropic pro-inflammatory cytokines, such as tumor necrosis factor $\alpha$ (TNF$\alpha$) and interleukin-1$\beta$ (IL-1$\beta$; Jung et al. 1995), which can act on neighboring epithelial cells and underlying immune cells. Furthermore, the intestinal epithelia can secrete PMN chemoattractants in a polarized manner so that PMNs are recruited to (and directed to migrate across) the intestinal epithelia. Specifically, in response to the pathogen *Salmonella typhimurium*, model intestinal epithelia secrete IL-8 basolaterally (McCormick et al. 1993, 1995) and pathogen-elicited epithelial chemoattractant (PEEC) apically (McCormick et al. 1998b). The IL-8 secretion serves to recruit PMN to a sub-epithelial compartment, and PEEC secretion drives the final step of PMN movement to the intestinal lumen. The fact that expression of these cytokines/chemokines is elevated in the intestinal mucosa of IBD patients (Mazzucchelli et al. 1994; Funakoshi et al. 1995; Daig et al. 1996; Nielsen et al. 1997) suggests that they are also involved in directing the movement of immune cells in these disorders. Thus, seeking to downregulate the secretion of such chemokines and/or seeking to downregulate neutrophil movement/activation in response to these chemokines is a very reasonable strategy for treatment of these disorders.

## 9.4 Lipoxins Are Natural Endogenously Biosynthesized Eicosanoids with Anti-Inflammatory Activity

Active inflammation in the intestine and elsewhere is normally self limiting. However, there is no known mechanism inherent in pro-inflammatory pathways that explains why this is the case. For example, considering the development scenario of inflammation described above, the inflammation would continue to amplify (i.e., cytokine secretion would lead to immune cell recruitment and more cytokine secretion, etc.) regardless of whether or not the infection (or other provoking agonist) was cleared. Thus, it has been postulated that anti-inflammatory pathways must also exist, although they have only recently begun to be recognized. One class of metabolites that appears to feature such endogenous anti-inflammatory mediators is the lipoxins (LXs; Fig. 1; Serhan 1997). LXs are lipoxygenase (LO)-interaction products of arachidonic acid. Their biosynthesis proceeds via the actions of two distinct LOs that are not expressed in any single cell type. Rather, LXs are made by the actions of more than one cell type on an individual molecule of arachidonic acid. For example, epithelial cells contain 15-LO, and neutrophils contain 5-LO. The actions of these enzymes enable neutrophils and epithelial cells that are in close proximity to cooperate to make LXs. A variety of other pathways convert the LO-interaction product into one or more of several possible species of LX. The requirement of cell cooperation for LX biosynthesis restricts LX production to already-inflamed sites; thus, these eicosanoids may play a role in limiting inflammation.

In contrast to most other arachidonate metabolites, which have bioactivities that promote inflammation, LX have an array of bioactivites (see below) that strongly suggest their in vivo role is that of counter-regulator for pro-inflammatory molecules. The best-characterized LX, both in terms of its bioactivity and mechanism of action, is lipoxin $A_4$ ($LXA_4$). This lipid mediator acts via a specific receptor (termed $LXA_4R$), a high affinity ($k_d \approx 1$ nM), seven-transmembrane $\alpha$-helix G-protein-coupled receptor (Fiore et al. 1994). Synthetic stable analogs of $LXA_4$ that are more resistant to degradation than the native eicosanoid yet retain the ability to specifically ligate $LXA_4R$ exhibit greater anti-inflammatory activity than $LXA_4$ (Serhan et al. 1995). One compound that behaves as an $LXA_4$-stable analog (in that it competes for the $LXA_4$ receptor, is

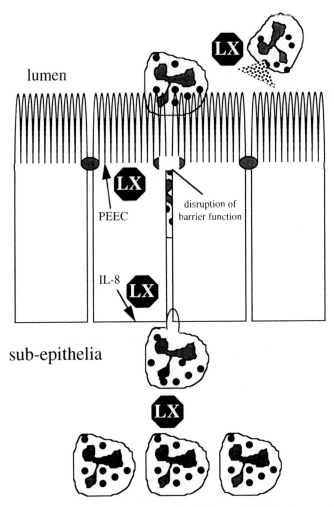

**Fig. 1.** Lipoxins (*LX*) act as *stop signals* for inflammation in both neutrophils and epithelial cells

resistant to enzymatic degradation and has anti-inflammatory bioactivity) is the 15-epimer of LXA4 (15-epi-LXA4; Claria and Serhan 1995). However, 15-epi-LXA4 is, in fact, biosynthesized in vivo by a pathway involving cyclo-oxygenase that has been acetylated by aspirin (Claria and Serhan 1995). Thus, some of the bioactions of aspirin may come from activation of counter-regulatory pathways via LXA4R.

## 9.5 Results from both In Vitro and In Vivo Studies Indicate that Lipoxin-Stable Analogs Likely Have Therapeutic Potential for Intestinal Diseases Characterized by Active Inflammation

In an in vitro model of intestinal inflammation, stable analogs of $LXA_4$ downregulate events that define and may cause intestinal inflammation. This bioaction is due to $LXA_4$ analogs that act both on the intestinal epithelia and on neutrophils, both of which express the $LXA_4R$ (Fiore et al. 1993; Gronert et al. 1998). $LXA_4$ and its analogs reduce neutrophil movement (Lee et al. 1989) and transepithelial migration (Colgan et al. 1993) to exogenous chemotactic gradients. Additionally, $LXA_4$ analogs downregulate epithelial-cell secretion of chemokines that direct neutrophil movement. Specifically, $LXA_4$ analogs reduce epithelial-cell IL-8 secretion by up to 80% in response to either the pro-inflammatory cytokine TNFα or the gastroenteritis-eliciting pathogen *S. typhimurium* (Gewirtz et al. 1998; Gronert et al. 1998). Furthermore, $LXA_4$ analogs downregulate epithelial secretion of PEEC by about 60% (Gewirtz et al. 1998). This downregulation of epithelial chemokine secretion reduces neutrophil movement through the underlying matrix and across intestinal epithelial monolayers. If, in this model system, we expose both the neutrophils and the epithelia to $LXA_4$ analogs, we observe nearly complete attenuation of neutrophil transepithelial movement (Gewirtz and Madara, unpublished observation). Since flare-ups of active disease in IBD patients may result from bacteria–epithelia–neutrophil interactions, this is a particularly encouraging observation. $LXA_4$-analog downregulation of neutrophil movement is likely mediated by the $LXA_4R$, as the ability of various $LXA_4$ analogs to exhibit this bioactivity correlates strictly with their ability to specifically ligate $LXA_4R$. Consistent with the $k_d$ of $LXA_4R$, $LXA_4$ and its analogs exhibit these bioactions at subnanomolar–nanomolar concentrations. The specific signaling mechanism by which $LXA_4$ analogs downregulate these pro-inflammatory functional events has not been defined but appears to be mediated via downregulation of nuclear factor κB (our unpublished results), a central switch in many pro-inflammatory pathways.

Not only are the symptoms of active IBD attributable (at least in large part) to neutrophil influx but, furthermore, the products these neutrophils secrete are likely responsible for much of the damage that is inflicted on host tissue during active intestinal inflammation. Specifically, both the reactive oxygen metabolites and granule proteases that activated neutrophils secrete can damage host tissue and can further promote the inflammatory state. $LXA_4$-stable analogs downregulate neutrophil release of these proteases, thus representing another potential mechanism by which these lipid mediators can be therapeutic for diseases characterized by active inflammation (Gewirtz et al. 1999). Importantly, such downregulation of neutrophil degranulation is observed in response to immune complexes similar to those found in abundance in several chronic inflammatory diseases, including IBD, adult respiratory-distress syndrome and glomerulonephritis. Like the above-described anti-inflammatory bioactions of $LXA_4$ analogs, $LXA_4$-analog downregulation of neutrophil degranulation is also mediated by $LXA_4R$. Interestingly, $LXA_4$ analogs do not reduce superoxide secretion in response to immune complexes but rather specifically downregulate the release of the granule population, which contains the proteases that can damage host tissue (i.e., azurophilic granules). This selective downregulation of a neutrophil-effector function could perhaps be a means of limiting damage to host tissue while maintaining the ability to kill ingested microorganisms via neutrophil oxidative burst (the products of which are much shorter lived than are granule proteases).

It seems likely (though it has not yet been tested) that these compounds will also be able to downregulate intestinal inflammation in vivo and thus may be potentially therapeutic for IBD. $LXA_4$ and its analogs potently downregulate inflammation in well-defined animal models, such as the hamster cheek pouch and the mouse ear (Raud et al. 1991; Takano et al. 1997). The anti-inflammatory bioactivity exhibited by $LXA_4$ analogs in this in vivo assay was more potent than that exhibited by equimolar concentrations of dexamethasone. It has recently been shown that human sections of colonic mucosa exhibit a downregulation of chemokine secretion (both IL-8 and macrophage chemotactic protein 1) in response to LXA$_4$-stable analogs (Goh 1999), supporting the notion that these analogs act in the intestine as they do in other tissues and in cell culture. In both this ex vivo system and the in vitro model described above, chemokine secretion was downregulated via treatment with LXA$_4$ analogs prior to the addition of a provoking agonist. Thus, we envision that these compounds will be able to downregulate the initiation of the active flares of IBD. We are currently testing the effects of LXA$_4$ analogs in mouse models of IBD.

## 9.6 Why LXA$_4$ Analogs, Unlike Non-Steroidal Anti-Inflammatory Drugs, are Unlikely to Exacerbate Intestinal Disease

Currently available treatments for IBD include steroids, general immune suppressors (cyclosporin) and aminosalicylates. The side effects of steroids and immunosupressors allow only short-term use of these agents. Aminosalicylates (which act by an unknown mechanism) are better tolerated and are used to maintain disease remission but are only marginally effective at doing so (Greenfield et al. 1993). One class of agents that has proven to be tremendously effective at treating inflammation in many other tissues but worsens chronic inflammatory diseases of the intestine are the non-steroidal anti-inflammatory drugs (NSAIDs). NSAIDs work, at least in part, by preventing prostaglandin (which is pro-inflammatory in most tissues) and thromboxane formation via cyclo-oxygenase. However, because prostaglandins (especially prostaglandin E$_2$) also play a role in maintenance of the epithelial barrier in the intestine, NSAIDs lead to ulcerations that cause or exacerbate disease (Hawkey 1996). While it has been suggested that the creation of NSAIDs that selectively target only the COX-2 isoform might spare intestinal tissue, recent studies in mice and rats indicate that this may not be the case (Reuter et al. 1996; Wallace et al. 1998a, 1998b). Thus, NSAIDs in the intestine seem to be "dual-edged swords" that do more harm than good. Since LXA$_4$-stable analogs exhibit anti-inflammatory bioactivity (largely by activating their receptor) and do not affect the activity of COX, they are unlikely to exacerbate intestinal inflammation in this manner. Even though aspirin leads to generation of stable ligands

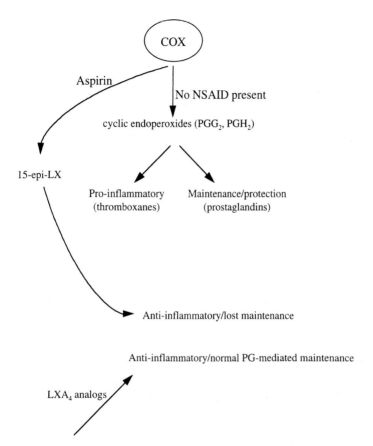

**Fig. 2.** Lipoxin A4 (*LXA4*) analogs should offer some of the anti-inflammatory effects of aspirin without any detrimental effects on the prostaglandin synthesis that maintains the well-being of the gastrointestinal tract

(15-epi-LX) for $LXA_4R$, like other NSAIDs it still blocks prostaglandin biosynthesis, likely accounting for its negative effects in the intestine. Therefore, direct administration of 15-epi-$LXA_4$ or $LXA_4$-stable analogs ought to provide the anti-inflammatory bioactivity that can be mediated by the $LXA_4R$ without the negative effects that occur when generating these compounds in vivo via aspirin treatment (Fig. 2).

## 9.7 Conclusions

In summary, $LXA_4$ analogs are a potentially very useful tool for downregulating inflammation in the intestine. They downregulate the defining and causative events of intestinal inflammation in an in vitro and ex vivo model. Further, these compounds potently downregulate inflammation (in other tissues) in animal models. These analogs work via a mechanism distinct from that used by current therapeutic agents (except, perhaps, part of aspirin's modus operandi; see above); thus, they may not encounter the problems experienced by those other agents. Especially exciting is the notion that instead of inhibiting enzymes or antagonizing receptors, $LXA_4$ analogs are probably activating the receptors by which inflammation is normally downregulated.

## References

Baumgart DC, McVay LD, Carding SR (1998) Mechanisms of immune cell-mediated tissue injury in inflammatory bowel disease. Int J Mol Med 1:315–32

Claria J, Serhan CN (1995) Aspirin triggers previously undescribed bioactive eicosanoids by human endothelial cell-leukocyte interactions. Proc Natl Acad Sci U S A 92:9475–9479

Colgan SP, Serhan CN, Parkos CA, Delp-Archer C, Madara JL (1993) Lipoxin A4 modulates transmigration of human neutrophils across intestinal epithelial monolayers. J Clin Invest 92:75–82

Daig R, Andus T, Aschenbrenner E, Falk W, Scholmerich J, Gross V (1996) Increased interleukin 8 expression in the colon mucosa of patients with inflammatory bowel disease. Gut 38:216–222

Fiore S, Romano M, Reardon E, Serhan CN (1993) Induction of functional lipoxin A4 receptors in HL-60 cells. Blood 81:3395–3403

Fiore S, Maddox JF, Perez HD, Serhan CN (1994) Identification of a human cDNA encoding a functional high affinity lipoxin A4 receptor. J Exp Med 180:253–260

Funakoshi K, Sugimura K, Sasakawa T, Bannai H, Anezaki K, Ishizuka K, Yoshida K, Narisawa R, Asakura H (1995) Study of cytokines in ulcerative colitis. J Gastroenterol 30[suppl]8:61–63

Gewirtz AT, McCormick B, Neish AS, Petasis NA, Gronert K, Serhan CN, Madara JL (1998) Pathogen-induced chemokine secretion from model intestinal epithelium is inhibited by lipoxin A4 analogs. J Clin Invest 101:1860–1869

Gewirtz AT, Fokin VV, Petasis NA, Serhan CN, Madara JL (1999) LXA$_4$, aspirin-triggered 15-epi LXA$_4$, and their stable analogs selectively down-regulate PMN azurophilic degranulation. Am J Physiol 276:C988–994

Goh J, Baird AW, Goodson C (1999) Stable lipoxin analogs attenuate chemokine release from cytokine activated human colonic epithelium. Gastroenterology (in press)

Greenfield SM, Punchard NA, Teare JP, Thompson RP (1993) Review article: the mode of action of the aminosalicylates in inflammatory bowel disease. Aliment Pharmacol Ther 7:369–383

Gronert KG, Gewirtz AT, Madara JL, Serhan CN (1998) Identification of a human enterocyte lipoxin A4 receptor that is regulated by IL-13 and INF-γ that inhibits TNF-α-induced IL-8 release. J Exp Med 187:1285–1294

Hawkey CJ (1996) Non-steroidal anti-inflammatory drug gastropathy: causes and treatment. Scand J Gastroenterol Suppl 220:124–127

Jung HC, Eckmann LY, Panja A, Fierer J, Morzycka-Wroblewska E, Kagnoff MF (1995) A distinct array of proinflammatory cytokines is expressed in human colon epithelial cells in response to bacterial invasion. J Clin Invest 95:55–65

Lee TH, Horton CE, Kyan-Aung U, Haskard D, Crea AE, Spur BW (1989) Lipoxin A4 and lipoxin B4 inhibit chemotactic responses of human neutrophils stimulated by leukotriene B4 and N-formyl-L-methionyl-L-leucyl-L-phenylalanine. Clin Sci 77:195–203

Madara J (1989) Loosening TJs. Lessons from the intestine. J Clin Invest 83:1089–1094

Madara JL, Nash S, Parkos C (1991) Neutrophil–epithelial cell interactions in the intestine. In: Wong P, Serhan C (eds) Cell–cell interactions and the release of inflammatory mediators. Plenum Press, New York

Madara JL, Parkos CA, Colgan SP, MacLeod RJ, Nash S, Matthews J, Delp C, Lencer WS (1992) Cl$^-$ secretion in a model intestinal epithelium induced by a neutrophil-derived secretagogue. J Clin Invest 89:1938–1944

Madara JL, Patapoff TW, Gillece-Castro B, Colgan SP, Parkos CA, Delp C, Mrsny RJ (1993) 5'-Adenosine monophosphate is the neutrophil-derived paracrine factor that elicits chloride secretion from T84 intestinal epithelial cells. J Clin Invest 91:2320–2325

Mazzucchelli L, Hauser C, Zgraggen K, Wagner H, Hess M, Laissue JA, Mueller C (1994) Expression of interleukin-8 gene in inflammatory bowel disease is related to the histological grade of active inflammation. Am J Pathol 144:997–1007

McCormick BA, Colgan SP, Archer CD, Miller SI, Madara JL (1993) *Salmonella typhimurium* attachment to human intestinal epithelial monolayers:

transcellular signalling to subepithelial neutrophils. J Cell Biol 123:895–907

McCormick BA, Hofman P, Kim J, Carnes D, Miller S, Madara J (1995) Surface attachment of *Salmonella typhimurium* to intestinal epithelia imprints the subepithelial matrix with gradients chemotactic for neutrophils. J Cell Biol 131:1599–1608

McCormick BA, Gewirtz A, Madara JL (1998a) Epithelial crosstalk with bacteria and immune cells. Curr Opin Gastroenterol 14:492–497

McCormick BA, Parkos CA, Colgan SP, Carnes DK, Madara JK (1998b) Apical secretion of a pathogen-elicited epithelial chemoattractant (PEEC) activity in response to surface colonization of intestinal epithelia by *Salmonella typhimurium*. J Immunol 160:455–456

Nielsen OH, Rudiger N, Gaustadnes M, Horn T (1997) Intestinal interleukin-8 concentration and gene expression in inflammatory bowel disease. Scand J Gastroenterol 32:1028–1034

Raud J, Palmertz U, Dahlen SE, Hedqvist P (1991) Lipoxins inhibit microvascular inflammatory actions of leukotriene B4. Adv Exp Med Biol 314:185–192

Reuter BK, Asfaha S, Buret A, Sharkey KA, Wallace JL (1996) Exacerbation of inflammation-associated colonic injury in rat through inhibition of cyclooxygenase-2. J Clin Invest 98:2076–2085

Serhan CN (1997) Lipoxins and novel aspirin-triggered 15-epi-lipoxins (ATL): a jungle of cell-cell interactions or a therapeutic opportunity? Prostaglandins 53:107–137

Serhan CN, Maddox JF, Petasis NA, Akritopoulou-Zanze I, Papayianni A, Brady HR, Colgan SP, Madara JR (1995) Design of lipoxin A4 stable analogs that block transmigration and adhesion of human neutrophils. Biochemistry 34:14609–14615

Takano T, Fiore S, Maddox JF, Brady HR, Petasis NA, Serhan CN (1997) Aspirin-triggered 15-epi-lipoxin A4 and LXA4 stable analogs are potent inhibitors of acute inflammation: evidence for anti-inflammatory receptors. J Exp Med 185:1693–1704

Wallace JL, Bak A, McKnight W, Asfaha S, Sharkey KA, MacNaughton WK (1998a) Cyclooxygenase 1 contributes to inflammatory responses in rats and mice: implications for gastrointestinal toxicity. Gastroenterology 115:101–109

Wallace JL, Reuter BK, McKnight W, Bak A (1998b) Selective inhibitors of cyclooxygenase-2: are they really effective, selective, and GI-safe? J Clin Gastroenterol 27[suppl]:S28–S34

Yamada T, Alpers DH, Owyang C, Powell DW, Silverstein FE (1995) Textbook of gastroenterology, 2nd edn. Lippincott, Philadelphia

Yardley JH (1986) Pathology of idiopathic inflammatory bowel disease and relevance of specific cell findings: an overview. In: Yardley JH (ed) Recent developments in the therapy of inflammatory bowel disease. Johns Hopkins, Baltimore, pp 3–9

# 10 The Role of Eicosanoids in Tumor Growth and Metastasis

D. Nie, K. Tang, K. Szekeres, M. Trikha, and K. V. Honn

10.1 Introduction .......................................... 201
10.2 Expression of 12-LOX in Cancers ...................... 203
10.3 12-LOX in Tumor Angiogenesis ........................ 204
10.4 Modulation of Apoptosis by 12-LOX ................... 206
10.5 12-LOX in Tumor Metastasis .......................... 208
10.6 12(S)-HETE as a Signaling Molecule .................. 209
10.7 Conclusion ........................................... 211
References ................................................ 212

## 10.1 Introduction

Mobilization of esterified arachidonic acid (AA) from membrane glycerolipid pools represents the key regulatory step in cellular responses to various stimuli, such as growth factors, cytokines, chemokines and circulating hormones. Released AA is metabolized via the cyclo-oxygenase (COX 1 and COX 2), lipoxygenase (LOX) or P450-dependent epoxygenase pathways to generate eicosanoids. In addition to their normal biological activities, such as stimulation of mitogenesis and cellular motility, eicosanoids have also been postulated to contribute to tumorigenesis and to the progression of certain tumor cells. Various AA metabolites have been implicated in a variety of growth-related signaling pathways involving ras (Han et al. 1991), interferon-α, epithelial growth factor (EGF), cyclic adenosine monophosphate, protein kinase C

(PKC; Hannigan and Williams 1991; Peppelenbosch et al. 1993; Tang et al. 1995a), mitogen-activated kinases (Rao et al. 1988) and fos (Danesch et al. 1994). Numerous studies have demonstrated a strong correlation between growth-factor-promoted cell proliferation and generation of various COX products, primarily prostaglandins (Nolan et al. 1988).

In the LOX family, three LOX pathways were originally identified, i.e., 5-LOX, 12-LOX and 15-LOX. However, recent data has blurred the line distinguishing 12-LOX from 15-LOX, as enzymes that contain both activities have been discovered. 5-LOX, 15-LOX and 12-LOX metabolize AA to the monohydroxy fatty acids 5(S)-hydroxyeicosatetraenoic acid (HETE), 15(S)-HETE and 12(S)-HETE, respectively. In addition, 15-LOX and some isoforms of 12-LOX can metabolize linoleic and other fatty acids to various bioactive lipids, including lipoxins and hepoxillins. 12-LOX exists as a family of isoforms. Originally, three major isoforms of 12-LOX were discovered. The first is the platelet-type 12-LOX normally expressed in platelets and recently found to be expressed in a variety of animal and human tumor cells. Platelet-type 12-LOX metabolizes only AA, to exclusively form 12(S)-HETE. The second isoform is leukocyte-type 12-LOX, which can metabolize both AA and linoleic acid to generate a mixture of 12(S)-HETE and 15(S)-HETE. The third isoform of 12-LOX, isolated from bovine tracheal epithelial cells and rat brain, shares significant homology with 15-LOX and leukocyte 12-LOX and less homology with platelet-type 12-LOX. Recently, two new isoforms of 12-LOX have been identified in mouse skin (Krieg et al. 1995).

Eicosanoids derived from LOX pathways of AA metabolism play an essential role in mediating growth-factor-stimulated cell growth. Examples include 15-HETE as a mitogenic regulator of T lymphocytes (Bailey et al. 1982), 12-HETE and leukotriene B4 as growth stimulators of epidermal cells (Chan et al. 1985), 12(S)-HETE stimulation of keratinocyte DNA synthesis (Kragballe and Fallon 1986) and 12- and 15-HETEs as mediators of insulin and EGF-stimulated mammary epithelial-cell proliferation (Bandopadhyay et al. 1988) and as synergistic effectors of basic-fibroblast-growth-factor- and platelet-derived growth factor (PDGF)-regulated growth of vascular endothelial cells and smooth-muscle cells (Dethlefsen et al. 1994). In this chapter, we want to focus on the role of 12-LOX and its major product, 12(S)-HETE, in the progression of cancers, with particular emphasis on prostate cancer.

## 10.2 Expression of 12-LOX in Cancers

### 10.2.1 Ectopic Expression of Platelet-Type 12-LOX in Cancers

There is compelling evidence to suggest that platelet-type 12-LOX is expressed in various tumor cells. 12-LOX messenger RNA (mRNA) have been detected in erythroleukemia, colon carcinoma, epidermoid carcinoma A431 cells, human glioma and prostate and breast cancer cells (Honn et al. 1994a). Rat and murine tumor cell lines also express 12-LOX (Chen et al. 1994; Hagmann et al. 1995). The sequences of reverse-transcriptase polymerase-chain-reaction 12-LOX products from human epidermoid A431 cells and human prostate-cancer cells and tissues have complete homology to platelet-type 12-LOX (Hagmann et al. 1995). The product of 12-LOX activity in tumor cells has been identified as predominantly the S enantiomer by chiral high-performance liquid chromatography, and its structure has been confirmed by gas chromatography–mass spectrometry spectral analysis (Liu et al. 1994b). In addition, in some cancer cell lines, 12-LOX mRNA has been found to be upregulated by cytokines, such as EGF and autocrine motility factor (Silletti et al. 1994).

### 10.2.2 Correlation of 12-LOX Expression with Tumor Stage in Prostate Carcinoma

In a study involving over 120 prostate-cancer patients, Gao et al. (1995) found that the level of 12-LOX mRNA expression is correlated with tumor stage. In this important clinical study, the expression level of 12-LOX and tumor stage, grade, positive surgical margins and lymph-node positivity were evaluated. Overall, 38% of 122 evaluable patients demonstrated levels of 12-LOX mRNA in prostate-cancer tissue that were elevated compared with the levels in matching normal tissues. An elevated level of 12-LOX was found in a significantly greater number of cases in T3, high-grade and surgical-margin-positive prostatic adenocarcinomas than in T2, intermediate- and low-grade and surgical-margin-negative prostatic adenocarcinomas. These data suggest that elevation of 12-LOX mRNA expression occurs more frequently in advanced stage, high-grade prostate cancer (Gao et al. 1995). These observations suggest that 12-LOX activity may be associated with prostate-cancer progression in vivo.

## 10.3 12-LOX in Tumor Angiogenesis

### 10.3.1 Angiogenesis

The growth of solid tumors requires persistent ingrowth of capillary vessels from pre-existing vasculature. Angiogenesis, the formation of new capillary networks, requires endothelial-cell proliferation, motility and tubular differentiation. The newly formed blood vessels not only deliver nutrients and oxygen to tumors for their continuous growth but also provide a gateway through which cancer cells can leave the primary sites and metastasize to distant organs. It is now well recognized that a rate-limiting step in solid-tumor growth is the recruitment of new capillary blood vessels from the host vasculature (Hanahan and Folkman 1996). The ability of a tumor to stimulate neovascularization is determined by its "angiogenic switch", whose on/off setting is dictated by the net balance of angiogenic stimulators and natural inhibitors (Hanahan and Folkman 1996) which, in turn, are determined by the balance between angiogenic factors and angiogenesis inhibitors.

### 10.3.2 Pro-Angiogenic Activity of 12(S)-HETE

The 12-LOX product 12(S)-HETE has been found to exert various effects on endothelial cells ranging from integrin surface expression to retraction. First, 12(S)-HETE upregulates the surface expression of integrin $\alpha_v\beta_3$. Tang et al. illustrated that 12(S)-HETE increases the surface expression of $\alpha_v\beta_3$ in both rat-aorta endothelial cells (Tang 1993a, 1993b) and murine pulmonary microvascular endothelial cells (CD3 and CD4; Tang et al. 1994, 1995b). It should be noted that integrin $\alpha_v\beta_3$ has been shown to be predominantly associated with angiogenic blood vessels in tumors and human wound granulation tissue (Brooks et al. 1994). Second, 12(S)-HETE can also induce endothelial-cell retraction. It has been shown that 12(S)-HETE induces a reversible, non-destructive, time- and dose-dependent retraction of endothelial cells (Honn et al. 1989). Furthermore, tumor cells can synthesize 12(S)-HETE in amounts sufficient to induce microvascular endothelial-cell retraction (Honn et al. 1994c). Co-incubation of Lewis lung-carcinoma cells or B16 amelanotic-cell melanoma (B16a) cells (but not 3T3 fibroblasts) with microvascular endothelial cells (CD3) resulted in a time-dependent retraction of the CD3 monolayers. Lewis lung-carcinoma-cell-induced endothelial-cell retraction was blocked by a specific LOX inhibitor, *N*-benzyl-*N*-hydroxy-5-phenylpentanamide (BHPP), but not by COX inhibitors (Honn et al. 1994c). Third, 12(S)-HETE is also a mitogen for microvasuclar endothelial cells (Tang et al. 1995a), especially at low concentrations of serum (Setty et al. 1987). In murine pulmonary microvascular endothelial cells (CD4), 12(S)-HETE enhanced growth and DNA synthesis in a time- and dose-dependent manner (Tang et al. 1995a). Furthermore, 12(S)-HETE promotes wound healing in injured CD4 endothelial-cell monolayers (Tang et al. 1995a).

### 10.3.3 Regulation of Prostate Cancer Angiogenesis by 12-LOX and 12(S)-HETE

Since platelet-type 12-LOX metabolizes AA to specifically form 12(S)-HETE, and since 12(S)-HETE, as described earlier, is an pro-angiogenic factor, we studied whether the expression of 12-LOX in human prostate carcinoma can influence the angiogenic phenotype of tumor cells. When expressed in human PCa cells, platelet-type 12-LOX had no detectable effect on tumor cell growth in vitro but strongly stimulated tumor growth in nude mice (Nie et al. 1998). The increased tumor growth was closely related to an increase in tumor vascularization. Furthermore, in a Matrigel implantation assay, tumor cells with high levels of 12-LOX expression were more angiogenic in vivo and secreted more motility factors to stimulate endothelial-cell migration in vitro, illustrating the fact that platelet-type 12-LOX increases the angiogenicity of prostate-carcinoma cells (Nie et al. 1998). The studies suggest that ectopic expression of platelet-type 12-LOX in PCa cells, either as an indirect result of genetic changes or due to other nutritional or environmental factors, may regulate the angiogenic potential of PCa cells.

## 10.4 Modulation of Apoptosis by 12-LOX

### 10.4.1 Apoptosis in Cancer

Apoptosis, or programmed cell death, is a genetically encoded cell-suicide program defined by characteristic morphologic, biochemical and molecular changes resulting in nonpathologic cell loss. A large number of distinct cellular phenotypes set apoptosis apart from another cell-death process, i.e., necrosis (Oltvai and Korsmeyer 1994; Majno and Joris 1995). Apoptosis plays a key role in physiological processes, such as embryonic development, maturation of the host immune system and maintenance of tissue and organ homeostasis. Apoptosis has also been implicated in a variety of pathological conditions exemplified by cardiac infarction, atherosclerosis, Alzeimer's disease and other neurodegenic diseases, human immunodeficiency virus, tumorigenesis and tumor progression. A multitude of factors have been implicated in regulating/modulating apoptosis. These include (Oltavi and Korsmeyer 1994; Ruoslahti and Reed 1994; Cory 1995; Hockenbery 1995; Korsemeyer et al. 1995; Kroemer et al. 1995; Majno and Joris 1995; Martin and Green 1995; Rubin and Baserga 1995):

1. Oncogenes/tumor suppressor genes exemplified by p53, the bcl-2 family (bcl-2, bcl-XL, bcl-Xb, bcl-XS, bax, BAG-1, bad, bak, A1, Mcl-1), myc, ras, abl, raf, Rb-1 and Waf-1
2. Growth factor/growth-factor receptors, represented by nerve-growth factor (NGF)/NGF receptor, tumor necrosis factor $\alpha$/Fas, transforming growth factor (TGF)/TGF receptor, insulin-like growth factor (IGF)-1/IGF receptor and PDGF/PDGF receptor
3. Intracellular signal transducers (such as PKC, tyrosine kinases and protein phosphatases), lipid signaling molecules (such as ceramide) and $Ca^{2+}$
4. Cell-cycle regulators exemplified by cdc-2 and E2F
5. Reactive oxygen species
6. Extracellular matrix (ECM) regulators/signal transducers (ECM proteins, such as fibronectin and transmembrane integrin receptors)
7. Specific endonucleases, such as $Ca^{2+}$- and $Mg^{2+}$-dependent deoxyribonuclease

8. Cytoplasmic proteases typified by the interleukin-1-converting enzyme family.

The major impact of apoptosis on cancer research is manifested primarily in three areas: oncogenesis, tumor homeostasis and the mechanism of action of cytotoxic antitumor drugs (Stewart 1994). Most antitumor agents, such as radiation (which generates oxygen radicals) and chemotherapeutic drugs, kill tumor cells by inducing apoptosis. Likewise, development of resistance to these treatments by tumor cells is mostly a result of loss of apoptosis (Stewart 1994; Majno and Joris 1995).

Apoptosis has been implicated in the pathogenesis of human prostate cancers and in the patient response to hormonal, chemo- and radiation therapy (Wheeler et al. 1994). Hormonal ablation is the mainstay of treatment for androgen-dependent prostate cancer. As tumor progression occurs, most PCa cells become androgen-independent and thus resistant to hormonal therapy. Elevated levels of bcl-2, an apoptosis-suppressing protein, have been associated with the generation of androgen-independent cell populations (McDonnel et al. 1992). Increased expression of bcl-2 and some other oncogenes (c-myc) may also be responsible for the resistance of PCa cells to various chemotherapeutic drugs (Sinha et al. 1995). These observations suggest that new therapeutic modalities, such as combined androgen ablation and cytotoxic (apoptosis-inducing) chemotherapy (Berchem et al. 1995) may be developed to treat both early and disseminated PCa patients.

### 10.4.2 Eicosanoid Regulation of Apoptosis

In addition to positively regulating cell growth, both COX and LOX products of AA metabolism may also be involved in modulating cell survival and apoptosis. Prostaglandins (PGE2, PGA2 and PGJ2) have been shown to induce apoptosis of tumor cells (Kim et al. 1993; Goetzel et al. 1995). By contrast, COX inhibitors (non-steroidal anti-inflammatory drugs) also trigger apoptosis of cultured fibroblasts (Lu et al. 1995), and overexpression of COX 2 confers resistance to apoptosis induction on cells (Tsujii and DuBois 1995). Similarly, LOX products, such as hydroperoxyeicosatetraenoic acid, have been proposed to be involved in

T-cell apoptosis in acquired-immune-deficiency-syndrome patients (Sandtrom et al. 1994). However, 5-LOX inhibitors can cause apoptosis of human prostate-cancer cells (Anderson et al. 1998).

### 10.4.3 Regulation of Apoptosis by 12-LOX

Our recent work has identified the arachidonate LOX pathway as a novel, physiologically important regulator of tumor-cell survival and apoptosis. Tang et al. (1996) demonstrated that 12-LOX (and, probably, 15-LOX) plays an essential role in apoptosis. Using a rat carcinosarcoma W256 cell line of monocytoid origin as the model, the authors reported that downregulation of 12-LOX gene expression by antisense oligonucleotides triggered time- and dose-dependent apoptosis. General inhibitors of the LOX activity and 12-LOX-selective inhibitors caused significant W256-cell death. These observations collectively suggest that AA metabolism plays a significant physiological role in regulating tumor-cell survival and apoptosis.

## 10.5 12-LOX in Tumor Metastasis

### 10.5.1 Tumor Metastasis

Metastasis is a process involving multiple interactions between tumor and host cells. For successful metastasis to occur, tumor cells must be able to detach from their primary site, intravasate, arrest at the endothelium, extravasate and be able to colonize the secondary site (Liotta 1986). These interactions involve tumor cell–platelet, tumor cell–endothelium and tumor cell–ECM interactions. Most cellular interactions are mediated by cell-adhesion molecules, such as cadherins, selectins, immunoglobulins and integrins. In addition to these glycoproteins, lectins, proteoglycans and sphingolipids also are known to participate in metastasis (Tang and Honn 1994; Varner and Cheresh 1996).

## 10.5.2 Modulation of Tumor Metastasis by 12(S)-HETE

In many experimental systems, there is a positive correlation among 12(S)-HETE biosynthesis, tumor cell adhesion, expression of integrin receptors and the metastatic potential of tumor cells. In the murine B16a model system, subpopulations of tumor cells with high metastatic potential can produce approximately fourfold more 12(S)-HETE than the low-metastatic-potential cells (Liu et al. 1994b). The high-metastatic-potential cells demonstrate a five- to tenfold increase in lung-colonizing ability, which was inhibited by the selective 12-LOX inhibitor BHPP. When high-metastatic-potential cells, but not low-metastatic-potential cells, adhered to an endothelial-cell monolayer, there was a surge in 12(S)-HETE production, which could be blocked by a 12-LOX inhibitor (BHPP; Liu et al. 1994b). In addition, in the rat Dunning-PCa model system, a positive correlation was found between metastatic ability in vivo and ability to generate 12(S)-HETE in vitro (Liu et al. 1994a). Finally, there are numerous examples that demonstrate that exogenous 12(S)-HETE can increase parameters related to the metastatic potential of tumor cells, i.e., increased motility (Timar et al. 1993), increased secretion of lysosomal proteinases cathepsin B and L (Honn et al. 1994b; Ulbricht et al. 1996), increased expression of integrin receptor $\alpha_{IIb}\beta_3$ (Timar et al. 1995), increased adhesion to endothelium (Timar et al. 1992), increased spreading on subendothelial matrix (Honn et al. 1989) and increased lung-colonizing ability in vivo (Liu et al. 1994b).

## 10.6 12(S)-HETE as a Signaling Molecule

The plethora of biological activities elicited by 12(S)-HETE suggests its activation of multiple cellular-signaling pathways. While a specific 12(S)-HETE receptor is yet to be identified, high-affinity binding sites on cell surfaces were described for B16a murine melanoma cells (Liu et al. 1995) and for keratinocytes (Arenberger et al. 1993). Low-affinity cell-surface binding sites were also found on peripheral blood mononuclear cells (Zakaroff-Girard et al. 1999). Furthermore, 12(S)-HETE-binding studies revealed a 650-kDa complex of hsp70, hsp90 and other unidentified proteins in the cytosol and in the nucleus; this complex had a high affinity for 12(S)-HETE (Herbertsson et al. 1998). The proposed cell-surface receptor is G-protein coupled, because 12(S)-HETE-stimulated signaling was able to be blocked by pertussis-toxin pretreatment (Szekeres et al., unpublished data). Early experiments demonstrated that 12(S)-HETE mimicked the phorbol ester phorbol myristate acetate in enhancing tumor-cell integrin expression and adhesion (Timar et al. 1995). Subsequently, it was demonstrated that 12(S)-HETE induced a 100% increase in membrane-associated PKCα activity (Liu et al. 1995). It was further demonstrated that the stimulation of tumor cells with 12(S)-HETE was followed by a rapid accumulation of diacyl glycerol and inositol-(1,4,5)-trisphosphate via activating an upstream G protein and phospholipase C (Liu et al. 1995). A current report suggests that 12(S)-HETE also stimulates activity of another phospholipase, phospholipase D (Zakaroff-Girard et al. 1999). Phospholipase products and an increase in cytosolic $[Ca^{2+}]$ (Yoshino et al. 1994) in turn promote membrane association of PKCα (Liu et al. 1995) and PKCδ (Hagerman et al. 1997).

In addition to its activation of the PKC pathway, 12(S)-HETE was shown to increase tyrosine phosphorylation of several proteins in focal adhesions, one of which was shown to be focal adhesion kinase (pp125FAK) and a 42/44-kDa doublet that migrated in the position of mitogen-activated protein kinase (MAPK; Tang et al. 1995c). Subsequently, it was demonstrated that 12(S)-HETE can increase phosphorylation of MAPK (Wen et al. 1996) and activate it (in part) through PKCα and Raf (Szekeres and Honn, unpublished data). Figure 1 summarizes the signaling events initiated by 12(S)-HETE. Recent findings indicate that 12(S)-HETE also activates phosphatidylinositide-3 kinase, which can activate MAPK in both PKC-dependent and -independent fashions (Fig. 1; Szekeres and Honn, unpublished data). In addition to p42/44 MAPK, 12(S)-HETE was found to mediate angiotensin-induced activation of another member of the MAPK family, c-Jun N-terminal kinase (Wen et al. 1997).

In various experimental settings, it was observed that the biological activities of 12(S)-HETE could be antagonized by 13(S)-hydroxyoctadecadienoic acid (HODE), a bioactive lipid derived from linoleic-acid metabolism by 15-LOX (Grossi et al. 1989; Tang et al. 1993a). The 13(S)-HODE activities appear to be mediated through inhibition of PKC activation by 12(S)-HETE. However, as in the case of 12(S)-HETE, 13(S)-HODE does not directly suppress PKC activation. 13(S)-HODE

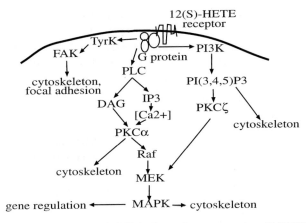

**Fig. 1.** Hypothetical model of 12(S)-hydroxyeicosatetraenoic acid (*HETE*) signal transduction. *DAG* diacyl glycerol, *FAK* focal adhesion kinase, *IP3* inositol-(1,4,5)-trisphosphate, *MAPK* mitogen-activated protein kinase, *PI3K* phosphatidylinositide-3-kinase, *PI(3,4,5)P3* phosphatidylinositol-(3,4,5)-trisphosphate, *PKC* protein kinase C, *PLC* phospholipase C, *TyrK* tyrosine kinase

does, however, block the translocation of PKC from the cytoplasm to the membrane, a transfer induced by 12(S)-HETE (Liu et al. 1995).

## 10.7 Conclusion

Since the first report describing the ectopic expression of 12-LOX in cancer in 1994, tremendous progress has been made regarding the functional role of this enzyme in tumor progression. It is now clear that increased 12-LOX expression renders tumor cells more angiogenic, more resistant to apoptosis and more invasive and metastasizing. Further elucidation regarding the way 12-LOX expression is upregulated and the way 12-LOX and its lipid product 12(S)-HETE regulate angiogenesis and tumor progression will surely provide more insights into the biology of cancers.

## References

Anderson KM, Seed T, Vos M, Mulshine J, Meng J, Alrefai W, Ou D, Harris JE (1998) 5-Lipoxygenase inhibitors reduce PC-3 cell proliferation and initiate non-necrotic cell death. Prostate 37:161–173

Arenberger P, Kemeny L, Ruzicka T (1993) Characterization of high-affinity 12(S)-hydroxyeicosatetraenoic acid (12(S)-HETE) binding sites on normal human keratinocytes. Epithelial Cell Biol 2:1–6

Bailey JM, Bryant RW, Low CE, Papillo MB, Vanderhoeck JY (1982) Regulation of T-lymphocyte mitogenesis by the leukocyte product 15-hydroxy-eicosatetraenoic acid (15-HETE). Cell Immunol 67:112–120

Bandyopadhyay GK, Imagawa W, Wallace DR, Nandi SJ (1988) Proliferative effects of insulin and epidermal growth factor on mouse mammary epithelial cells in primary culture. Enhancement by hydroxyeicosatetraenoic acids and synergism with prostaglandin E2. J Biol Chem 263:7567–7573

Berchem GJ, Bosseler M, Sugars LY, Voeller HJ, Zeitlin S, Gelmann EP (1995) Androgen induce resistance to bcl-2-mediated apoptosis in LNCaP prostate cancer cells. Cancer Res 55:735–738

Brooks PC, Clark RAF, Cheresh DA (1994) Requirement of vascular integrin $\alpha_v\beta_3$ for angiogenesis. Science 264:569–572

Chan C, Duhamel E, Ford-Hutchinson A (1985) Leukotriene B4 and 12-hydroxyeicosatetraenoic acid stimulate epidermal proliferation in vivo in the guinea pig. J Invest Dermatol 85:333–334

Chen YQ, Duniec ZM, Liu B, Hagmann W, Gao X, Shimoji K, Marnett LJ, Johnson CR, Honn KV (1994) Endogenous 12(S)-HETE production by tumor cells and its role in metastasis. Cancer Res 54:1574–1579

Cory S (1995) Regulation of lymphocyte survival by the bcl-2 gene family. Annu Rev Immunol 13:513–543

Danesch U, Weber PC, Sellmayer A (1994) Arachidonic acid increases c-fos and Egr-1 mRNA in 3T3 fibroblasts by formation of prostaglandin E2 and activation of protein kinase C. J Biol Chem 269:27258–27263

Dethlefsen SM, Shepro D, D'Amore PA (1994) Arachidonic acid metabolites in bFGF-, PDGF-, and serum-stimulated vascular cell growth. Exp Cell Res 212:262–271

Gao X, Grignon DJ, Chbihi T, Zacharek A, Chen YQ, Sakr W, Porter AT, Crissman JD, Pontes JE, Powell IJ, Honn KV (1995) Elevated 12-lipoxygenase mRNA expression correlates with advanced stage and poor differentiation of human prostate cancer. Urology 46:227–237

Goetzel EJ, An S, Smith WJ (1995) Specificity of expression and effects of eicosanoid mediators in normal physiology and human diseases. FASEB J 9:1051–1058

Grossi IM, Fitzgerald LA, Umbarger LA, Honn KV (1989) Bidirectional control of membrane expression and/or activation of the tumor cell IRGpIIb/IIIa receptor and tumor cell adhesion by lipoxygenase products of arachidonic acid and linoleic acid. Cancer Res 49:1029–1039

Hagerman RA, Fischer SM, Locniskar MF (1997) Effect of 12-O-tetradecanoylphorbol-13-acetate on inhibition of expression of keratin 1 mRNA in mouse keratinocytes mimicked by 12(S)-hydroxyeicosatetraenoic acid. Mol Carcinog 19:157–64

Hagmann W, Gao X, Zacharek A, Wojciechowski LA, Honn KV (1995) 12-Lipoxygenase in Lewis lung carcinoma cells: molecular identity, intracellular distribution of activity and protein, and Ca(2+)-dependent translocation from cytosol to membranes. Prostaglandins 49:49–62

Han JW, McCormick F, Macara IG (1991) Regulation of Ras-GAP and the neurofibromatosis-1 gene product by eicosanoids. Science 252:576–579

Hanahan D, Folkman J (1996) Patterns and emerging mechanisms of the angiogenic switch during tumorigenesis. Cell 86:353–364

Hannigan GE, Williams BR (1991) Signal transduction by interferon-alpha through arachidonic acid metabolism. Science 251:204–207

Herbertsson H, Kuhme T, Evertsson U, Wigren J, Hammarstrom S (1998) Identification of subunits of the 650 kDa 12(S)-HETE binding complex in carcinoma cells. J Lipid Res 39:237–244

Hockenbery DM (1995) Bcl-2, a novel regulator of cell death. Bioessays 17:631–638

Honn KV, Grossi IM, Diglio CA, Wojtukiewicz M, Taylor JD (1989) Enhanced tumor cell adhesion to the subendothelial matrix resulting from 12(S)-HETE induced endothelial cell retraction. FASEB J 3:2285–2293

Honn KV, Tang DG, Gao X, Butovich IA, Liu B, Timar J, Hagmann W (1994a) 12-lipoxygenases and 12(s)-HETE: role in cancer metastasis. Cancer Metastasis Rev 13:365–396

Honn KV, Timár J, Rozhin J, Bazaz R, Sameni M, Ziegler G, Sloane BF (1994b) A lipoxygenase metabolite, 12(S)-HETE, stimulates protein kinase C-mediated release of cathepsin B from malignant cells. Exp Cell Res 214:120–130

Honn KV, Tang DG, Grossi I, Duniec ZM, Timar J, Renaud C, Leithauser M, Blair I, Johnson CR, Diglio CA, Kimler VA, Taylor JD, Marnett LJ (1994c) Tumor cell-derived 12(S)-hydroxyeicosatetraenoic acid induces microvascular endothelial cell retraction. Cancer Res 54:565–574

Kim IK, Lee JH, Sohn HW, Kim HS, Kim SH (1993) Prostaglandin A2 and δ2-prostaglandin J2 induce apoptosis in L1210 cells. FEBS Lett 321:209–214

Korsmeyer SJ, Yin X-M, Oltvai ZN, Veis-Novack DJ, Linette GP (1995) Reactive oxygen species and the regulation of cell death by the Bcl-2 gene family. Biochim Biophys Acta 1271:63–66

Kragballe K, Fallon JD (1986) Increased aggregation and arachidonic acid transformation by psoriatic platelets: evidence that platelet-derived 12-hydroxy-eicosatetraenoic acid increases keratinocyte DNA synthesis in vitro. Arch Dermatol Res 278:449–453

Krieg P, Kinzig A, Ress-Loschke M, Vogel S, Vanlandingham B, Stephan M, Lehmann WD, Marks F, Furstenberger G (1995) 12-Lipoxygenase isoenzymes in mouse skin tumor development. Mol Carcinog 14:118–128

Kroemer G, Petit P, Zamzami N, Vayssiere J, Mignotte B (1995) The biochemistry of programmed cell death. FASEB J 9:1277

Liotta LA (1986) Tumor invasion and metastasis – role of the extracellular matrix: Rhoads Memorial Award lecture. Cancer Res 46:1–7

Liu B, Maher RJ, Hannun YA, Porter AT, Honn KV (1994a) 12(S)-HETE enhancement of prostate tumor cell invasion: selective role of PKC$\alpha$. J Natl Cancer Inst 86:1145–1151

Liu B, Marnett LJ, Chaudhary A, Chuan J, Blair IA, Johnson CR, Diglio CA, Honn KV (1994b) Biosynthesis of 12(S)-hydroxyeicosatetraenoic acid by B16 amelanotic melanoma cells is a determinant of their metastatic potential. Lab Invest 70:314–323

Liu B, Khan WA, Hannun YA, Timar J, Taylor JD, Lundy S, Butovich I, Honn KV (1995) 12(S)-hydroxyeicosatetraenoic acid and 13(S)-hydroxyoctadecadienoic acid regulation of protein kinase C-$\alpha$ in melanoma cells: role of receptor-mediated hydrolysis of inositol phospholipids. Proc Natl Acad Sci U S A 92:9323–9327

Lu X, Xie W, Reed D, Bradshaw WS, Simmons DL (1995) Nonsteroidal antiinflammatory drugs cause apoptosis and induce cyclooxygenases in chicken embryo fibroblasts. Proc Natl Acad Scid U S A 92:7961–7965

Majno G, Joris I (1995) Apoptosis, oncosis, and necrosis. An overview of cell death. Am J Pathol 146:3–15

Martin SJ, Green DR (1995) Protease activation during apoptosis: death by a thousand cuts? Cell 82:349–352

McDonnel TJ, Troncoso P, Brisbay SM, Logothetis C, Chung LWK, Hsieh J, Tu S, Campbell ML (1992) Expression of the protooncogene bcl-2 in the prostate and its association with emergence of androgen-independent prostate cancer. Cancer Res 52:6940–6944

Nie D, Hillman GG, Geddes T, Tang K, Pierson C, Grignon DJ, Honn KV (1998) Platelet-type 12-lipoxygenase in a human prostate carcinoma stimulates angiogenesis and tumor growth. Cancer Res 58:4047

Nolan RD, Danilowicz RM, Eling TE (1988) Role of arachidonic acid metabolism in the mitogenic response of BALB/c 3T3 fibroblasts to epidermal growth factor. Mol Pharmacol 33:650-656

Oltvai ZN, Korsmeyer S (1994) Checkpoints of dueling dimers foil death wishes. J Cell Biol 79:189-192

Peppelenbosch MP, Tertoolen LGJ, Hage WJ, de Laat SW (1993) Epidermal growth factor-induced actin remodeling is regulated by 5-lipoxygenase and cyclooxygenase products. Cell 74:565-575

Rao GN, Baas AS, Glasgow WC, Eling TE, Runge MS, Alexander WR (1988) Activation of mitogen-activated protein kinases by arachidonic acid and its metabolites in vascular smooth muscle cells. J Biol Chem 269:32586-32591

Rubin R, Baserga R (1995) Insulin-like growth factor-I receptor. Its role in proliferation, apoptosis, and tumorigenicity. Lab Invest 73:311-331

Ruoslahti E, Reed J (1994) Anchorage dependence, integrins, and apoptosis. Cell 77:477-478

Sandtrom PA, Tebbey PW, Van Cleave S, Buttke TM (1994) Lipid hydroperoxides induce apoptosis in T cells displaying a HIV-associated glutathione peroxidase deficiency. J Biol Chem 269:798-801

Setty BN, Graeber JE, Stuart MJ (1987) The mitogenic effect of 15- and 12-hydroxyeicosatetraenoic acid on endothelial cells may be mediated via diacylglycerol kinase inhibition. J Biol Chem 262:17613-17622

Silletti S, Timar J, Honn KV, Raz A (1994) Autocrine motility factor induces differential 12-lipoxygenase expression and activity in high- and low-metastatic K1735 melanoma cell variants. Cancer Res 54:5752-5756

Sinha BK, Yamazaki H, Eliot HM, Schneider E, Borner MM, O'Connor PM (1995) Relationships between proto-oncogene expression and apoptosis induced by anticancer drugs in human prostate tumor cells. Biochim Biophys Acta 1270:12-18

Stewart BW (1994) Integration of genetic, biochemical, and cellular indicators. J Natl Cancer Inst 86:1286-1296

Tang DG, Honn KV (1994) 12-lipoxygenases, 12(S)-HETE, and cancer metastasis. Ann N Y Acad Sci 744:199-215

Tang DG, Chen YQ, Renaud C, Diglio CA, Honn KV (1993a) Protein kinase C-dependent effects of 12(S)-HETE on EC vitronectin receptor and fibronectin receptor. J Cell Biol 121:689-704

Tang DG, Diglio CA, Honn KV (1993b) 12(S)-HETE-induced microvascular endothelial cell retraction results from PKC-dependent rearrangement of cytoskeletal elements and $\alpha_v\beta_3$ integrins. Prostaglandins 45:249-268

Tang DG, Diglio CA, Honn KV (1994) Activation of microvascular endothelium by 12(S)-HETE leads to enhanced tumor cell adhesion via upregulation of surface expression of $\alpha_v\beta_3$ integrin: a post-transcriptional, PKC- and cytoskeleton-dependent process. Cancer Res 54:1119–1129

Tang DG, Renaud C, Stojakovic S, Diglio CA, Porter A, Honn KV (1995a) 12(S)-HETE is a mitogenic factor for microvascular endothelial cells: its potential role in angiogenesis. Biochem Biophys Res Commun 211:462–468

Tang DG, Chen YQ, Diglio CA, Honn KV (1995b) Transcriptional activation of endothelial cell integrin $\alpha_v$ by protein kinase C activator 12(S)-HETE. J Cell Sci 108:2629–2644

Tang DG, Grossi IM, Tang KQ, Diglio CA, Honn KV (1995c) Inhibition of TPA and 12(S)-HETE-simulated tumor cell adhesion by prostacyclin and its stable analogs: rationale for their antimetastatic effects. Int J Cancer 60:418–425

Tang DG, Chen YQ, Honn KV (1996) Arachidonate lipoxygenases as essential regulators of cell survival and apoptosis. Proc Natl Acad Sci U S A 93:5241–5246

Timar J, Chen YQ, Liu B, Bazaz R, Taylor JD, Honn KV (1992) The lipoxygenase metabolite 12(S)-HETE promotes cytoadhesion molecule $\alpha_{IIb}\beta_3$-mediated tumor cell spreading on fibronectin. Int J Cancer 52:594–603

Timar J, Silletti S, Bazaz R, Honn KV (1993) Regulation of melanoma-cell motility by the lipoxygenase metabolite 12(S)-HETE. Int J Cancer 55:1003–1010

Timar J, Bazaz R, Kimler V, Haddad M, Tang DG, Robertson D, Tovari J, Taylor JD, Honn KV (1995) Immunomorphological characterization and effects of 12-(S)-HETE on a dynamic intracellular pool of the $\alpha_{IIb}\beta_3$ integrin in melanoma cells. J Cell Sci 108:2175–2186

Tsujii M, DuBois R (1995) Alterations in cellular adhesion and apoptosis in epithelial cells overexpressing prostaglandin endoperoxide synthase 2. Cell 83:493–501

Ulbricht B, Hagmann W, Ebert W, Spiess E (1996) Differential secretion of cathepsins B and L from normal and tumor human lung cells stimulated by 12(S)-hydroxy-eicosatetraenoic acid. Exp Cell Res 226:255–263

Varner JA, Cheresh DA (1996) Integrins and cancer. Curr Opin Cell Biol 8:724–730

Wen Y, Nadler JL, Gonzales N, Scott S, Clauser E, Natarajan R (1996) Mechanisms of ANG II-induced mitogenic responses: role of 12-lipoxygenase and biphasic MAP kinase. Am J Physiol 271:C1212–C1220

Wen Y, Scott S, Liu Y, Gonzales N, Nadler JL (1997) Evidence that angiotensin II and lipoxygenase products activate c-Jun NH2-terminal kinase. Circ Res 81:651–655

Wheeler TM, Rogers E, Aihara M, Scardino PT, Thompson TC (1994) Apoptotic index as a biomarker in prostatic intraepithelial neoplasia (PIN) and prostate cancer. J Cell Biochem Suppl 19:202–207

Yoshino H, Kitayama S, Morita K, Uchiyama Y, Shibata K, Shirakawa M, Okamoto H, Tsujimoto A, Dohi T (1994) Effect of 12-hydroxyeicosatetraenoic acid on cytosolic calcium in human neutrophils. J Lipid Mediat Cell Signal 9:225–234

Zakaroff-Girard A, Gilbert M, Meskini N, Nemoz G, Lagarde M, and Prigent A-F (1999) The priming effect of 12(S)-hydroxyeicosatetraenoic acid on lymphocyte phospholipase D involves specific binding sites. Life Sci 64:2135–2148

# Subject Index

active-site structure of LTA$_4$ hydrolase  89
airway hyper-responsiveness  174
ALXR  162
ALXR mRNA  163
Alzheimer's disease  54
aminopeptidase  86
angiogenesis  204
angioplasty  175
arachidonic acid  4
arthritis  174
aspirin  1, 9, 168
aspirin initiated ATL-biosynthesis pathways  173
aspirin-sensitive asthmatics  11
asthma  174
atherosclerotic  175
Atlantic salmon  168

bestatin  92
bi-directional transcellular biosynthesis  156
biological half-life  148
blast crisis  174

C20:4  174
captopril  92
catalytic mechanisms  89
CD11b/CD18  169
celecoxib  14
chemokine receptors  132

chiral HPLC  59
chloride stimulation  87
colon cancers  6, 13
core temperature  112
COX  4, 10, 29, 56, 65, 145
COX-1  97
COX-2  97
COX immunostaining  116
COX isoforms  98
COX null cells  104
Cys-LT1  136
cytoplasmic tail  132
cytoprotective  6
cytosolic phospholipase A$_2$  125

13,14-dihydro-15-oxo-LXA$_4$  177
13,14-dihydro-LXA$_4$  177

endothelial cells  7, 118
endothelial dysfunction  172
15-epi-LXA$_4$  172
epithelia  188
epithelial growth factor  157
epoxide hydrolase  86
etodolac  10, 11

familial adenomatous polyposis  13
fever  7, 98
fibroblast growth factor  157
Fiji nuts  155
fish  155, 171

frogs 171

G-protein 132
G-protein-coupled receptor 128
glia-like cells 118
glomerulonephritic animals 176
glomerulonephritis 174
GTPase activity 159

heme 56
HL-60 cell 128, 177
HUVEC 163
15-hydroxyprostaglandin dehydrogenase 177

IL-8 189, 192
IL-8 secretion 166, 189
ileal mesentery 169
inflamed glomeruli 167
inflammation 187, 192, 196
inhibitors 92
intestine 187, 194

12-keto-LTB$_4$ 127
knockout mice 98
– phenotypes 171

leukotrienes 85, 144
lipopolysaccharide 5
lipoxins 143
5-lipoxygenase 125
15-LO
– transgenic overexpression 171
LPS 7, 98
LTA$_4$ 85
– hydrolase 127
LX analogs 166
LX-biosynthetic enzymes 157
LXA$_4$ 169
– receptor 159

meloxicam 10, 11

membrane proteins 54
mesangial cells 163
microcirculation 170
mouse ALXR 162
MPA kinase 136
MS–MS spectra 154
mutational analysis 88

N-glycosylation 162
nasal polyps 174
neutrophil 188, 192
nimesulide 10, 11
nitric-oxide generation 166
non-steroid anti-inflammatory drugs (NSAIDs) 2, 8, 65, 99

Onchorynchus mykiss 168
osteoarthritis 11
overlapping active sites 89
15-oxo-LXA$_4$ 177

P-selectin 172
P-selectin-dependent interactions 173
peritoneum 173
phagocytosis 168
16-phenoxy-LXA$_4$ 172
phospholipase A$_2$ 25
platelet-derived growth factor 157
platelet-PMN interactions 171
PMA 104
PMN recruitment 167
pneumonia 174
prostacyclin 5, 8
prostaglandin endoperoxide H synthase 53
prostaglandins 5, 10, 13, 65, 144
PTX-insensitive manner 164
purinergic receptor 131

rabbit aorta-contracting substance 3

## Subject Index

rat models  169
replication  53
rheumatoid arthritis  2, 11
15(R/S)-methyl-LXA$_4$  172

S$_N$1 mechanism  91
salicylic acid  1
sarcoidosis  174
subtraction strategy  128
suicide inactivation  89

thermolysin  88
THP-1 cells  164

thrombosis  54
transcellular eicosanoid biosynthesis  147
tyrosyl radical  56

vascular injury  172
vasodilatory properties  170

whole-blood assays  9, 10

zinc metallohydrolases  86
zinc site in LTA$_4$ hydrolase  86

Ernst Schering Research Foundation Workshop

*Editors:* Günter Stock
Monika Lessl

*Vol. 1 (1991):* Bioscience ⇆ Society – Workshop Report
*Editors:* D. J. Roy, B. E. Wynne, R. W. Old

*Vol. 2 (1991):* Round Table Discussion on Bioscience ⇆ Society
*Editor:* J. J. Cherfas

*Vol. 3 (1991):* Excitatory Amino Acids and Second Messenger Systems
*Editors:* V. I. Teichberg, L. Turski

*Vol. 4 (1992):* Spermatogenesis – Fertilization – Contraception
*Editors:* E. Nieschlag, U.-F. Habenicht

*Vol. 5 (1992):* Sex Steroids and the Cardiovascular System
*Editors:* P. Ramwell, G. Rubanyi, E. Schillinger

*Vol. 6 (1993):* Transgenic Animals as Model Systems for Human Diseases
*Editors:* E. F. Wagner, F. Theuring

*Vol. 7 (1993):* Basic Mechanisms Controlling Term and Preterm Birth
*Editors:* K. Chwalisz, R. E. Garfield

*Vol. 8 (1994):* Health Care 2010
*Editors:* C. Bezold, K. Knabner

*Vol. 9 (1994):* Sex Steroids and Bone
*Editors:* R. Ziegler, J. Pfeilschifter, M. Bräutigam

*Vol. 10 (1994):* Nongenotoxic Carcinogenesis
*Editors:* A. Cockburn, L. Smith

*Vol. 11 (1994):* Cell Culture in Pharmaceutical Research
*Editors:* N. E. Fusenig, H. Graf

*Vol. 12 (1994):* Interactions Between Adjuvants, Agrochemical
and Target Organisms
*Editors:* P. J. Holloway, R. T. Rees, D. Stock

*Vol. 13 (1994):* Assessment of the Use of Single Cytochrome
P450 Enzymes in Drug Research
*Editors:* M. R. Waterman, M. Hildebrand

*Vol. 14 (1995):* Apoptosis in Hormone-Dependent Cancers
*Editors:* M. Tenniswood, H. Michna

*Vol. 15 (1995):* Computer Aided Drug Design in Industrial Research
*Editors:* E. C. Herrmann, R. Franke

*Vol. 16 (1995)*: Organ-Selective Actions of Steroid Hormones
*Editors:* D. T. Baird, G. Schütz, R. Krattenmacher

*Vol. 17 (1996)*: Alzheimer's Disease
*Editors:* J.D. Turner, K. Beyreuther, F. Theuring

*Vol. 18 (1997):* The Endometrium as a Target for Contraception
*Editors:* H.M. Beier, M.J.K. Harper, K. Chwalisz

*Vol. 19 (1997)*: EGF Receptor in Tumor Growth and Progression
*Editors:* R. B. Lichtner, R. N. Harkins

*Vol. 20 (1997)*: Cellular Therapy
*Editors:* H. Wekerle, H. Graf, J.D. Turner

*Vol. 21 (1997)*: Nitric Oxide, Cytochromes P 450,
and Sexual Steroid Hormones
*Editors:* J.R. Lancaster, J.F. Parkinson

*Vol. 22 (1997)*: Impact of Molecular Biology
and New Technical Developments in Diagnostic Imaging
*Editors:* W. Semmler, M. Schwaiger

*Vol. 23 (1998)*: Excitatory Amino Acids
*Editors:* P.H. Seeburg, I. Bresink, L. Turski

*Vol. 24 (1998)*: Molecular Basis of Sex Hormone Receptor Function
*Editors:* H. Gronemeyer, U. Fuhrmann, K. Parczyk

*Vol. 25 (1998):* Novel Approaches to Treatment of Osteoporosis
*Editors:* R.G.G. Russell, T.M. Skerry, U. Kollenkirchen

*Vol. 26 (1998):* Recent Trends in Molecular Recognition
*Editors:* F. Diederich, H. Künzer

*Vol. 27 (1998):* Gene Therapy
*Editors:* R.E. Sobol, K.J. Scanlon, E. Nestaas

*Vol. 28 (1999):* Therapeutic Angiogenesis
*Editors:* J.A. Dormandy, W.P. Dole, G.M. Rubanyi

*Vol. 29 (2000):* Of Fish, Fly, Worm and Man
*Editors:* C. Nüsslein-Volhard, J. Krätzschmar

*Vol. 31 (2000):* Advances in Eicosanoid Reseach
*Editors:* C.N. Serhan, H.D. Perez


*Supplement 1 (1994):* Molecular and Cellular Endocrinology of the Testis
*Editors:* G. Verhoeven, U.-F. Habenicht

*Supplement 2 (1997)*: Signal Transduction in Testicular Cells
*Editors:* V. Hansson, F. O. Levy, K. Taskén

*Supplement 3 (1998)*: Testicular Function:
From Gene Expression to Genetic Manipulation
*Editors:* M. Stefanini et al.

*Supplement 4 (2000)*: Hormone Replacement Therapy
and Osteoporosis
*Editors:* J. Kato, H. Minaguchi, Y. Nishino

*Supplement 5 (1999):* Interferon:
The Dawn of Recombinant Protein Drugs
*Editors:* J. Lindenmann, W.D. Schleuning